T0292851

Projects: Methods: Outcomes

Matching the speed of change in modern business, this book takes readers on a two-year journey in building a project management office (PMO) for today and tomorrow and redefines the PMO as to what it should focus on: Projects, Methods, and Outcomes.

Many organisations invest heavily in PMOs, but these are built on an outdated and static model that does not fit a hybrid, agile, AI-empowered, and rapidly changing business environment. Building on his renowned "balanced PMO" model, project management leader Peter Taylor tackles today's challenges with this diary-style guide to inspire all PMO leaders, project managers, and business leaders, and provide a road map to follow to build (or rebuild) their own PMOs. He presents a completely new definition of "PMO", eliminating the traditional back-office concept of a centralised PMO, with his "Projects: Methods: Outcomes" construct that provides a truly business-focused team to oversee the delivery of value to their organisation.

Enriched with case studies and practical models, this book will benefit all PMO leaders, project management professionals, change and transformation leaders, and anyone interested in how to deliver business value through projects.

Keynote speaker and coach, **Peter** is the author of the number-one best-selling project management book "The Lazy Project Manager", along with many other books on Project Management.

He has built and led some of the largest PMOs in the world with organisations such as Siemens, IBM, UKG, and now Ceridian (Dayforce), where he is the VP of Global PMO. He has also delivered over 500 lectures around the world in 26 countries and has been described as "perhaps the most entertaining and inspiring speaker in the project management world today".

You can discover more about Peter through his website: www.thelazyprojectmanager.com.

Projects: Methods: Outcomes

The New PMO Model for True Project and Change Success

Peter Taylor

Routledge
Taylor & Francis Group

NEW YORK AND LONDON

Designed cover image: Getty

First published 2024
by Routledge
605 Third Avenue, New York, NY 10158

and by Routledge
4 Park Square, Milton Park, Abingdon, Oxon, OX14 4RN

Routledge is an imprint of the Taylor & Francis Group, an informa business

ISBN: 978-1-032-38732-1 (hbk)
ISBN: 978-1-032-38730-7 (pbk)
ISBN: 978-1-003-34647-0 (ebk)

DOI: 10.4324/9781003346470

Typeset in Sabon
by MPS Limited, Dehradun

As this is, most likely, my last book relating to the world of project management and PMOs[1] (although never say never), then I dedicate this to everyone who has ever helped me on my long and wonderful project journey, and also to all those who follow in my path building and leading project communities all over the world.

Thank you and good luck!

Peter

Note

1   Which is why I am certain that you will forgive me for quoting from (nearly) all my books throughout this volume.

Contents

Acknowledgements

As this book is the culmination of a 40-year career and the summation of many books that I have had the pleasure of writing and the joy of hearing from (mostly[2]) happy readers, the list of acknowledgements and "thank-you" notes is a very, very long one, so name-checking everyone would be longer than this actual book.

Instead, please accept the following as my sincere appreciation.

Those key to the making of this book and building the best PMO I have ever had the honour to lead are listed in full in the Appendices, to which I add the team at Routledge who has helped in the creation of this publication.

Beyond that, if our paths have crossed over the years, whether as part of the project communities I have been a small part of, or through my speaking (in many wonderful countries) and writing (from blogs and articles through to the books), or just in general, then I truly appreciate your support and help.

The journey is not yet over for me, and I would be delighted to come and visit some new places and meet some new people to "talk" all things project (just contact me and make the invitation[3]).

And finally, to my wife, Juliet,[4] who knows that if the office door is closed and Metallica[5] is playing, then I am writing, and it's probably best if I am left alone – thank you.[6]

Notes

2   You can't please all the people is a truism – "You can please some of the people all of the time, you can please all of the people some of the time, but you can't please all of the people all of the time". John Lydgate
3   www.thelazyprojectmanager.com.

4   https://juliet-taylor.com/.
5   Metallica is an American heavy metal band formed in 1981 in Los Angeles by vocalist and guitarist James Hetfield and drummer Lars Ulrich. The authors' go-to song to "get creative" is "For Whom the Bells Toll" followed by "Enter Sandman".
6   And, of course, the "anyone else who knows me" catch-all.

Foreword

I've had the pleasure of watching Peter Taylor work his magic on all things project management and organisational transformation for several years – whether on the main stage at global conferences, 1:1 with customers, or helping enterprise technology organisations overhaul and modernise their project methods and tools.

He brings to his craft an enviable combination of vision, real-world experience, and results-based solutions.

And in his latest book, Peter again leans forward and has produced a comprehensive guide that challenges traditional project management practices and redefines the role of the project management office (PMO) in driving success.

Recognising the need for a new approach to project management, Peter, a renowned figure in the field, joins forces with his colleagues to introduce us to a ground-breaking methodology in his book titled "Projects: Methods: Outcomes – The new PMO model for true project and change success".

In the expanding landscape of project management and organisational transformation, it is essential to have a reliable, effective, and progressive framework that leads to true project and change success. With the rapid pace of technological advancements, shifting market dynamics, and increasing customer demands, organisations need to adapt and evolve their project management approaches to stay competitive and achieve their goals.

Peter's extensive experience and deep insights into project management have made him a trusted advisor for numerous organisations striving to deliver successful projects amidst complexity. In "Projects: Methods: Outcomes", he goes beyond the conventional PMO models that may have become outdated or ineffective in the face of contemporary challenges. Instead, he presents a fresh perspective that centres around three essential elements: Projects, Methods, and Outcomes.

By taking readers on a practical journey spanning 24 months, Peter and his colleagues provide invaluable insights into building a Global PMO overseeing hundreds of projects each year. The book offers an honest account of their experiences, accompanied by a wealth of tips, tricks, and strategies for achieving your own PMO success. Whether you are an experienced project manager, a change management professional, or a PMO leader, this book serves as a treasure trove of knowledge and inspiration.

"Projects: Methods: Outcomes – The new PMO model for true project and change success" equips readers with the tools and guidance necessary to navigate the complexities of project delivery in today's ever-evolving business landscape. Peter's expertise shines through his engaging writing style, making the book an invaluable resource for individuals seeking to achieve project success and drive meaningful change within their organisations.

With this comprehensive guide at hand, project managers, change agents, and PMO leaders will gain the necessary insights and strategies to transform their approaches, align with emerging trends, and ultimately achieve true project and change success. By embracing the principles outlined in "Projects: Methods: Outcomes", organisations can position themselves as industry leaders, driving innovation and achieving remarkable outcomes in an increasingly competitive market.

Steve Holdridge: President, Global Customer and Revenue Operations – Ceridian Inc.

Steve leads Ceridian's global field operations, including sales, customer success, services, partner, and support functions, with a focus on delivering customer business results and a best-in-class customer experience at every touchpoint.

Introduction

Most organisations invest heavily in project management, programme management, and portfolio management, and to manage all of this they invest in a PMO to oversee all matters of a "project". But these are typically built on a model that was designed ten plus years ago and has not significantly moved with the new hybrid, business agile, AI empowered, rapid speed of change demands of today's businesses.

This book takes readers on a practical journey of 24 months of designing and then building a PMO for today and for tomorrow, focusing on "Projects", "Methods", and "Outcomes".

Building on the "Balanced PMO" model I developed back in 2010 and detailed in the book "Leading Successful PMOs", but also focusing on the real needs of today, this book will be a diary-style guide for all PMO leaders, project managers, and business leaders to gain inspiration from, as well as an inspirational roadmap to follow to build or rebuild their own PMOs.

Case studies and practical models to follow guide the reader to build or improve their own PMO. It is definitely a "dip in and out" sort of book, but also follows a logical flow of PMO construction and purpose, not a methodology but an evolutionary path.

And be clear, when this book talks about "PMOs", it is truly talking about "big PMO". My PMO is a global PMO reaching across the world, overseeing thousands of projects each year. So big stuff, big scope, and big challenges (and ambition) – but take what applies to your PMO world and adapt as needed. It is all good stuff! And I believe it is all scalable to your own PMO needs.

But so, what? you might ask. Another book on PMOs – why? There are hundreds out there already (go check online yourself; I got bored after page 5). Why does the world need yet another one?

Well, the key to this book is a completely new definition of what a "PMO" means.

DOI: 10.4324/9781003346470-1

Traditionally, thoughts would go to "Projects", or "Programme", or "Portfolio" Management Office, but that creates a back-office PMO concept of a centralised PMO when in fact the PMO should be ultimately responsible for the outcomes of changes sanctioned by their businesses.

The "Projects: Methods: Outcomes" construct completely changes this attitude and provides a real-business focused team to oversee the delivery of value to their organisation.

I feel this is what PMOs have been missing in the past and I am seeing the benefits of this model every day as my own PMO grows in stature and scales in delivery.

Join me on the journey of the first two years of building this specific global PMO.

What you will get

Part 1 of the book sets the stage by exploring the historical context of PMOs and providing a reality check for organisations seeking to achieve true project and change success. It unveils the true meaning of a PMO, emphasising the importance of aligning projects, methods, and desired outcomes.

In Part 2, we take a journey through the process of building a roadmap for the PMO within your organisation. Sharing invaluable recommendations and key findings, emphasising the significance of a professional project management community, delivery assurance, and the need to address change fatigue. By adopting the balanced PMO model, organisations can effectively focus their efforts on delivering the desired outcomes while maintaining a high level of project management excellence.

Part 3 delves into the critical aspects of branding, marketing, engagement, and communication. We highlight the value and purpose of branding the PMO and its services and emphasise the role of effective communication in driving value for the organisation. By applying the "Stop/Start" principle, organisations can optimise their project management practices and achieve better results.

Part 4 focuses on delivering the PMO roadmap, providing practical advice on building a strong project community, implementing a flexible project framework, establishing a project academy, and leveraging AI to explore the future of project management. He also emphasises the importance of team building and communication, both for productivity and employee satisfaction.

Success and celebration are the themes of Part 5. The discussion moves on to measuring progress, celebrating achievements, and the significance of continuous learning. By embracing a culture of learning and celebrating milestones, organisations can foster a positive environment that encourages excellence and innovation.

Part 6 shares experiences in launching the PMO and provides insights into creating successful events and regional delivery experiences, prompting us to question whether an "event" is necessary for our organisations and provides guidance on making informed decisions.

Adoption and change, two inevitable challenges in any organisational transformation, are addressed in Part 7. Here, we delve into the difficulties of change and adoption, offering strategies to build a portfolio dashboard, drive key performance indicators (KPIs), and win the hearts and minds of stakeholders.

Part 8 offers reflections from the author's extensive journey in project management. He shares key insights on time management, the importance of sponsorship, and the enduring commitment required to make a difference.

Finally, Part 9 concludes the book with a collection of best PMO tips, providing practical advice and actionable strategies for PMO leaders and practitioners.

And then you have the Appendices, packed full of loads of extra ideas and guidance, and more go-to information.

"Projects: Methods: Outcomes – The new PMO model for true project and change success" is a treasure trove of knowledge and wisdom for anyone involved in project management, organisational change, or PMO leadership. Peter Taylor's expertise, combined with his engaging writing style, makes this book an invaluable resource for navigating the complexities of project delivery and achieving success in an ever-evolving business landscape.

What you won't get

To be fully transparent, this is not a methodology to build a successful PMO type of book.

It is not a step-by-step guide to PMO construction, and it is not a PMO in a box type of offering either.

So, if that is what you are looking for, then you need to go elsewhere, sorry. Hopefully, you kept your receipt, and you will be able to get your money back.

But if you are seeking suggested areas of critical importance that you should be considering, and if you are wanting inspiring examples of building an amazing PMO team and an excellent range of PMO services, then keep on reading. Money well spent.

Because you are in right place.

Ducks in row

The choice of cover image for this book was made with a sense of considering the leadership and management of a PMO – requiring that many "ducks" are needed to be aligned in some way in order to make any real progress and to bring home to roost the objectives (yes, there will be a number of terrible duck-related references now).

To have your ducks in a row means making sure all the pieces are in place before diving into a new venture. Now, there are a few theories on how this phrase waddled into existence.

One tale takes us to the early history of bowling when the bowling pins were chunkier and squatter, earning them the nickname "ducks". Before fancy machines did the job, someone had to manually line up these "duck pins" after each round. So, having your ducks in a row was like having those bowling pins all neat and tidy before rolling your next ball.

Another theory takes flight in the animal kingdom. Mother ducks, the multitasking marvels that they are, corral their adorable offspring into orderly lines before journeying over land or water. By keeping them in line, any escapees or stragglers would be immediately noticeable. Just like a mother duck, getting your ideas, tasks, or team members in a (reasonably) straight line would resemble the organised row of literal ducks. Yes, I know, baby ducks are called ducklings so maybe this theory isn't the right one since no one says, "getting your ducklings in a row".

But wait, there's more! Some sources argue that our ducks in a row saying owes its origins to the carnival game where a player takes aim with a small rifle or air gun to knock down moving targets. Often, these targets resemble ducks, and a conveyor belt system ensures they're lined up like a feathered parade. So, the expression might have sprung from the convenience of having all the targets (ducks) arrive in a predictable and organised order (like life is like that …).

And finally, let's soar high in the sky with a final theory. When ducks fly together, they form a V-formation, with one leader leading the way and the others following suit. This formation minimises wind resistance and allows each duck to ride the slipstream. Having all your metaphorical ducks in a row is as efficient and logical as those ducks flying in a well-organised flock.

So, whether you're lining up bowling pins, herding ducks, aiming at carnival targets, taking flight in formation, or just leading the complex world of a PMO, having your ducks in a row ensures you're ready to succeed.

The challenge is, and here is a warning, the fact you have your ducks lined up perfectly does not mean that they are lined up how any, or all of your stakeholders, want them to be lined up.

That is why leading a PMO is such a great (but challenging) job.

The other reason I chose this image is I just like colourful plastic ducks.

Part 1

The true meaning of "PMO"

In Part 1, we kick off with important background information before undertaking a reality check to ground us before we explore what is meant by the true meaning of "PMO" as a collective term.

DOI: 10.4324/9781003346470-2

Chapter 1.1

Background

A short history lesson

In 2011, I authored a popular book[1] on PMOs called "Leading Successful PMOs: How to Build the Best Project Management Office for Your Business". In it, I defined the concept of a PMO as the department or group responsible for setting and maintaining process standards related to project management within a company.

However, I also observed that the acronym "PMO" was totally confusing in that it can stand for Programme Management Office, Portfolio Management Office, Project Office, Project Control Office, Central Project Office, or Project Support Office, and more.

I argued that the PMO's goal is to standardise project execution and introduce economies of repetition, providing documentation, guidance, and metrics on project management practices. It also serves as the link between business strategy and the projects required to implement it.

Since the time of writing that book, I am delighted to see that businesses worldwide are increasingly assigning PMOs to exert overall influence and facilitate continuous organisational improvement by defining, borrowing, and collecting best practices in process and project management.

Interestingly, at the time of writing, I struggled to find 100 recognisable PMOs worldwide – today, the number is overwhelming, which is (mostly) a good thing.

In my follow-up book, "Delivering Successful PMOs: How to design and deliver the best project management office for your business", I emphasised that a PMO is an androgynous term that can be built to serve different purposes. Thus, its functions must be clearly defined and uniformly perceived within the organisation. I also declared that each and every PMO is (and must be) unique to the organisation that they serve.

While the fundamentals of PMOs remain the same all these years later, there have been significant changes since I first wrote about them. The rise of virtual work (pre- and post-pandemic), VUCA,[2] AI's[3] transformational

DOI: 10.4324/9781003346470-3

impact, and the new work/life balance drive have all made PMOs more complex.

Having had the opportunity to help (re)build a global PMO for a billion-dollar HCM[4] organisation starting in 2021, I share my team's journey in this book, presenting their thoughts and ideas. I hope to provide readers with words of encouragement, guidance, and inspiration, along with "best PMO" tip offerings for further consideration.

I encourage readers to take what they wish from our journey and lessons learned and create their own unique PMO, as each organisation's PMO must evolve as the business itself evolves.

Reality check

Before we get into it, then it is well worth thinking about why PMOs have been troubled in the past and are still challenged even today.

PMOs can face a variety of challenges, including:

1 Lack of executive support: PMOs require executive support to be effective. Without support from senior leadership, it can be difficult for PMOs to get the necessary resources, authority, and buy-in from stakeholders.
2 Inadequate funding and resources: PMOs require adequate funding and resources to deliver value to the organisation. Without sufficient resources, PMOs may struggle to establish the necessary standards, processes, and governance.
3 Resistance to change: One of the biggest challenges is resistance to change, which can come from various sources, including stakeholders, sponsors, project managers, and team members.
4 Lack of clarity on roles and responsibilities: The roles and responsibilities of PMOs can be unclear, leading to confusion and frustration among team members.
5 Resistance to standardisation: PMOs aim to standardise processes and procedures, but this can sometimes be met with resistance from team members who prefer their own methods.
6 Limited visibility and control: PMOs can struggle to maintain visibility and control over all projects and activities, especially in large organisations with many different departments and stakeholders.
7 Keeping up with technology: PMOs must stay up to date with the latest project management tools and technologies, but this can be a challenge, especially in rapidly evolving industries.
8 Balancing flexibility and standardisation: PMOs must balance the need for standardisation with the need for flexibility to accommodate different project types, sizes, and complexities.

Thinking about my own (current) PMO, I can honestly say that points 1 and 2 (which are real PMO "killers") have not been an issue for myself and my team. The business has strongly invested in and supported this particular PMO.[5] For this, I am truly grateful.

Points 3 to 8 we had to deal with, just like all other PMOs, and this book will touch on each and every one of these challenges, together with some ways in which my team and I have addressed them.

Notes

1   Well, I would argue that it was, and remains, popular due to the fact that it has been in and out of the Amazon project management book charts for 13 years now.
2   VUCA stands for volatility, uncertainty, complexity, and ambiguity. It describes the situation of constant, unpredictable change that is now the norm in certain industries and areas of the business world. VUCA demands that you avoid traditional, outdated approaches to management and leadership, and day-to-day working.
3   Artificial intelligence (AI) is intelligence – perceiving, synthesising, and inferring information – demonstrated by machines as opposed to intelligence displayed by humans.
4   Human Capital Management.
5   In fact, the foreword of this book is written by one of these sponsors – Steve Holdridge, and the chapter on Adoption is written by another – David Ayling-Smith.

Chapter 1.2

The true meaning of "PMO"

Projects, Methods, and Outcomes

It came at the very start of my current PMO journey, as I designed a roadmap to meet the needs and the vision of my organisation.

What does PMO actually stand for?

Amazing to have a globally recognised acronym but equally amazingly frustrating to have no actual simple definition, hard and fast, of what "PMO" stands for.

Linking this in one of those moments of unexpected inspiration together with a common problem I had encountered personally and had also seen in other PMOs suddenly gave me insight into the answer.

The heart of all PMOs worked well: the machinery behind the scenes if you like, the engine room, the back office. Methodology, training, standards, and so on, all good.

But the connection with the "front office", the "coal-face", whatever term you wish to use, was often less than solid. The engagement with the project managers and change agents had, as one of my team leaders loves to state, "room for improvement".

And so we have architected our PMO into three sub-teams, working in an integrated manner – Projects, Methods, and Outcomes (Figure 1.2.1).

Actually, what happened was my newly formed team challenged me to state what I thought "PMO" stood for, and I realised that they had asked an excellent question.[1]

Simon Sinek, a well-known author and speaker, had this to say about asking questions:[2]

"It's likely that if you're having trouble understanding something, then somebody else in the room is as well. Asking questions doesn't mean you're the stupidest person in the room; it usually means you're the only one brave enough to speak up".

This is a very fair point and most likely one that my PMO team was experiencing at the time.

DOI: 10.4324/9781003346470-4

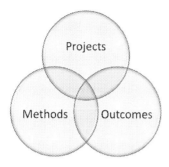

Figure 1.2.1 The "new" PMO model.

Harvard Business Review had this to say further about questions:[3]

"Questioning is a uniquely powerful tool for unlocking value in organizations: It spurs learning and the exchange of ideas, it fuels innovation and performance improvement, it builds rapport and trust among team members. And it can mitigate business risk by uncovering unforeseen pitfalls and hazards".

And so, I am really thankful to my team who asked this question, who challenged me to simply explain what I meant by the term *PMO*. It made me think, it made me reflect, and it let me conclude (I believe correctly) that in our world, "PMO" stands for "Projects | Methods | Outcomes".

Let's start with Projects.

- Projects were all about onboarding, education, certification, support, community, and the project manager career.
- Methods were all about the "how", the framework for common project delivery whilst offering flexibility on approach depending on project scale, partnership, and service offering.

This brings us to the "O" on PMO: Outcomes.

- Outcomes were the direct project management and service management interaction and support. The bridge between the Projects and Methods Team and the users of the output from those teams. A proactive two-way communications channels to share PMO strategy and to listen to, and react to, local tactical needs.

And there you have it.

I, and my team, stand behind this model of "Projects: Methods: Outcomes", which provides a real-business-focused team to oversee the delivery of value to their organisation.

Hence, the title of this book, "Projects: Methods: Outcomes – The new PMO model for true project and change success".

Notes

1   The Value of a great question was explored in one podcast episode from "The Squid of Despair" https://www.buzzsprout.com/1917876/episodes/10245051.
2   Sinek, Simon. "The Truth about Being the 'Stupidest' in the Room I Simon Sinek". www.youtube.com/watch?v=BkLzo_oNVho.
3   "How to Ask Great Questions". *Harvard Business Review*, May 2018.

Part 2

Building a roadmap for your organisation

In Part 2, we will begin with the roadmap to success and the building of the best team possible to deliver that success.

We will explore the Balanced PMO model and the new "Projects: Methods: Outcomes" model in more detail, whilst also covering the important strategies of "KISS" and "MURA", both of which are intrinsically linked to the author's long-held view of the value of "productive laziness".

DOI: 10.4324/9781003346470-5

Building a PMO roadmap for your organisation

Begin at the beginning, where else?

What was the purpose of the PMO, and therefore what was the PMO roadmap intended to deliver?

Before becoming the full-time leader of the Global PMO, I was engaged in a two-month consulting activity to produce:

1 A PMO roadmap report
2 A supporting executive briefing (document and presentation)
3 An action plan for achieving the global PMO roadmap

Vision statement

Driving this review was the overarching organisational strategy that translated into the following GPMO strategy.

To develop and present a global PMO roadmap aligning and supporting the key initiatives as part of the "10X Vision":

1 Global: Scale, Coverage, Capability
2 Delivery Innovation and Excellence: Quality, Value, Speed
3 Profitable Growth: Revenue, Margin, Value
4 Expanded Eco-System: SI/Partner Success
5 Enterprise: Capability and Customer Experience
6 Employee Experience: Engagement and Excellence

Identifying a growth path from the current "supportive" PMO position to one where the PMO is proactively at the heart of all client project engagements and operates at a global enterprise level with a focus on services, but partners with support, product, and sales as appropriate.

DOI: 10.4324/9781003346470-6

Executive summary

The following are the highest-level recommendations that I made to move the existing project management community "from tactical task masters to deliverers of vision"

To achieve this, the global PMO:

- Needs to be independent, strong, authoritative, objective, and leading.
- The central source of truth and content for all client project-based matters, guiding a global community of collaborative, mature PMs.
- Operating a disciplined PMO model acting as a (suitable) enforcement agency.
- Offering leadership, best practice, innovations, and future state initiatives, with Ceridian clients at its heart.

Unlearning

It was also noted that one of the first challenges to be faced was the need to "unlearn" (see Appendices for more).

Based on the "we have done it this way forever" attitude apparent in some areas, a process of "unlearning" will be required to re-align mindsets to the "right" way of working and leading project engagements.

The global PMO will have to have a clear understanding of this "unlearning" process and mindset change.

Being Brilliant

Rather than looking at a PMO Maturity Model for the PMO evolution, I felt that a four-stage growth plan better suited the organisation's needs and provided better clarity for progress measurement.

Be Brilliant at Basics

- This is about getting that framework for project delivery in place that addresses all forms of the project.
- It is about strengthening the project community to build a truly collaborative community that supports each other, as well as putting into practice the Project Academy model.
- And it is about designing, building, and launching all of the delivery assurance services.

Be Brilliant at Global

- This is about a regional PMO presence, together with cultural alignment and the optimisation of offshore resources.

Be Brilliant at Scale

- This is about the ability to consistently deliver at scale anywhere in the world – directly from our own resources, or in alignment with partners.

Be Brilliant at Futures

- This is about gaining from many actual and anticipated developments in the project world (AI being perhaps the most disruptive).

One of the key changes made was the renaming of the "old" PMO to the Global PMO. This was in direct response to the need for the PMO team to think "global" in everything that they did and to support the overall re-branding of the team to the new vision and purpose.

Chapter 2.2

Key findings and recommendations

The following summarises the key findings and recommendations that the report submitted and that were, in turn, approved by the executive team.

Methodology

The objective is to have one point of reference for project delivery; one source of the "truth".

- There is a requirement for the development of a framework that supports all types of projects.
- Such a framework must be fully scalable across all projects, globally.
- Quality of artifacts to be much improved (and look professional and consistent) with client collaborative elements incorporated.
- The PMO should be overseeing field training on the outcomes of this initiative and the "new" methodology, as well as leading adherence validation.

Professional project management community

The objective is to make our company a highly respected place for career project managers to be, in order to attract, develop, and retain the best in the market.

- Project management's overall "maturity" of behaviour, business awareness, responsibility, team leadership (purpose), and client interaction need to be elevated for global consistency.
- The creation of a Ceridian Project "Academy" – with a defined curriculum and measured grades/success through a balance of internal and external components – would give a clear definition to this PM career path.

DOI: 10.4324/9781003346470-7

Delivery assurance

The objective is to develop a formal process required for project assurance, health checks, retrospectives, and overall delivery assurance.

Linked to this, a standard escalation path/process is needed, led by the PMO, to provide rapid, focused support for client project issues and to relieve the pressure on senior management.

- Development and delivery of project "health" services
  - Health checks
  - Retrospectives
  - Escalation management
  - Interventions
  - Mentoring
- Learning from mistakes (close the cycle) must all be part of the daily process of projects and PMO initiatives and services.

Going global

The objective is to move the PMO away from a North American focused unit to a global and client-facing unit, which is critical in order to support the ambitions of growth and global presence.

- This means that the PMO must include regional representation and have the right level of current experience in practical project delivery from PMO team members.
- Segment representatives require the time to appropriately represent both their segments and the PMO, which right now is impacted by client project pressures and priorities.
- Offshore resources need to be optimised for both project delivery and PMO membership.

Change fatigue

It was noted strongly by the project managers (who were surveyed as part of the report preparation) that the impact of so many small and seemingly constant process changes was overwhelming, and they were experiencing change fatigue.

The objective was to design a mechanism to coordinate changes in process, method, reporting, and action, limiting "releases" to the field through periodic updates rather than the current "a little and often" approach.

It is also necessary to field test changes prior to general release to "prove" effectiveness and value.

Adoption

All the above should increase the levels of adoption and adherence to such changes and gather real-world insights from the field.

Chapter 2.3

Your world

> Best PMO Tip[1]: Consider getting some objective help in developing your own PMO roadmap, whether you are starting from nothing or, more likely, you are doing a rebuild.

Naturally, your PMO and project world will be different in some or more ways. Every PMO should be different, and every business requires a different slant, a different priority (or priorities[2]), and a different focus.

Your task is to undertake a review and to develop a suitable roadmap that your executives could sign off on and that your organisation needs.

The approach I took consisted of a large number of interviews with management and leading project community members, as well as a survey of all project managers globally. Follow-up summaries of findings were shared and discussed with key leaders before the full report was produced.

This was then presented to an executive team for consideration and, I am happy to say, accepted for the most part.

At this point, I was "sold" on the company, the vision, the opportunity, and the culture and therefore, when asked, I was more than happy to sign up for a full-time role as VP Global PMO.

The rest is, as they say, history (and fully documented in this book as well).

Happy days.

Notes

1 Throughout this book, you will see these "Best PMO Tips" that I hope will allow you to focus on the critical aspects of your own PMO. They are all summarised in Part 9 for easy reference.

DOI: 10.4324/9781003346470-8

2 "The word priority came into the English language in the 1400s". It was
   singular. It meant the very first or prior thing. It stayed singular for the next
   five hundred years. Only in the 1900s did we pluralise the term and start
   talking about priorities. Illogically, we reasoned that by changing the word we
   could bend reality. Somehow, we would now be able to have multiple "first"
   things. Greg McKeown, Essentialism.

The balanced PMO

How the model works and how to focus

Best PMO Tip: The best PMOs balance all of this to achieve the most effective development of capability, representation of capability, and sharing of capability and achievement.

For reference, this is a summary from my book "Leading Successful PMOs", where I first laid out the concept and model for a balanced PMO.

As with most things in life (and business), getting a balance right can prove far more effective, especially in the long run, than having a single focus that ignores other key elements.

The same is true of the PMO.

A balanced approach will definitely pay dividends and will not only ensure that the PMO is as effective and efficient as possible, but will also aid the acceptance of the PMO by the rest of the organisation.

For example, if your PMO is created solely with the purpose of being the "project police", then you will be in for a very short run. No doubt the role of policing projects is one part of the PMO's responsibility, but not the only part; such an approach may work for a short period of time, but it is not sustainable. And if your PMO is focused on firefighting, then again it will work for a while, but not beyond a certain point as it is demoralising to only work on problem projects and deal with escalating issues. It is far better to prevent the fires from even starting.

One way to achieve such a balance in the PMO is to consider structuring your efforts under what I call the "5 Ps":

• P = People
• P = Process
• P = Promotion

DOI: 10.4324/9781003346470-9

- P = Performance
- P = Project Management Information System

It may be tempting to just think of the PMO as all about the process, the means to ensure that good project management is achieved through methodology and quality assurance, etc., but that ignores the people side.

And it may be that your consideration is towards the project management community and your focus is drawn towards the people (projects are all about people after all), and so you direct your efforts as a PMO leader towards training and team building, etc., but this ignores the project mechanics.

You may also accept the need to build a good tracking and reporting system, supported by an investment in a project management information system, to deliver the visibility of project health and progress towards business goals.

But without the inclusion of a promotional programme, it could well be the case that all of the good work you and your team achieve in the areas of process and people will go unnoticed and unappreciated by both your peers and the executive (see the "Marketing is your Friend" chapter for more thoughts in this area) (Figure 2.4.1).

Above is an example of a "5 Ps" model for one organisation (yours will be different), just as a reference point.

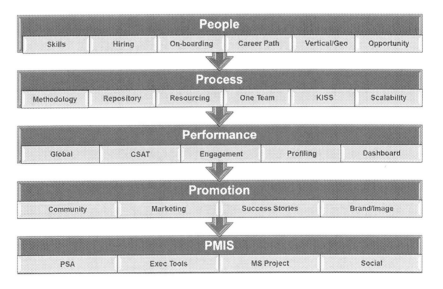

Figure 2.4.1 Example of a 5 Ps model.

Another way to look at what your PMO should be doing is as follows. What does the "P" stand for in "PMO", over and above the usual suspects? Well, how about

- The Process Management Office

It is about method and process, isn't it?

Or maybe

- The Performance Management Office

How are the projects doing? The project managers? Hell, even the PMO, how is that delivering?

And also

- The People Management Office

People in the PMO, people linked to the PMO, and PMO stakeholders, how are they being looked after by your PMO?

The key is to think big and wide and inclusively when designing the PMO and identifying your goals and objectives.

What makes your PMO of value to the business?

The new PMO model

The Balanced PMO model has served me well across a number of organisations, but there was something missing and it was when I joined my
latest company and sat down with my PMO leaders to brainstorm the
future path that this "missing link" was identified and filled (as was
presented earlier in "The true meaning of 'PMO'").

Best PMO Tip: The best PMOs will consider investing in the
"outcomes" focus for their future PMO model.

Projects | Methods | Outcomes

- Projects Team focused on onboarding, education, certification, support, community, and the project manager career.
- Methods Team focused on the "how", the framework for common
  project delivery whilst offering flexibility on approach, depending on
  project scale, partnership, and service offering.
- Outcomes Team focused on being the direct project management and
  service management interaction and support. The bridge between the
  Projects and Methods Teams and the users of the output from those
  teams. A proactive, two-way communications channel to share the
  PMO strategy and to listen to, and react to, local tactical needs.

And the three teams, under the banner of the "Global PMO" working
together to support the business at both a tactical level and at a strategic level
(sometimes quite a challenging and conflicting position it has to be said).

In all honesty, this took a while to get up to running in a smooth and
integrated way. Creating three teams naturally leads to three independent
views of priority and I note in all fairness that we have been through a
tough learning curve on this.

DOI: 10.4324/9781003346470-10

Just consider the Tuckman[1] model of team building – Forming, Storming, Norming, and Performing – and multiply that by three for the three teams and then again at a full Global PMO level. A deeper look at this can be found in my book "Team Analytics: The Future of High-Performance Teams and Project Success".[2]

What I would say here is that the teams and the team leaders have made this transition a success and, I am delighted to say, have identified and self-managed a number of these integration issues and the Global PMO is now a highly effective working unit.

Notes

1   The forming–storming–norming–performing model of group development was first proposed by Bruce Tuckman, in 1965, who said that these phases are all necessary and inevitable in order for a team to grow, face up to challenges, tackle problems, find solutions, plan work, and deliver results.
2   Team Analytics: The Future of High-Performance Teams and Project Success – Goldman, B and Taylor, P – Publisher Routledge 2023.

Creating the best PMO team to deliver the vision

> Best PMO Tip: The best PMOs have the very best people as part of the PMO team and ensure that they have a mix of project/programme knowledge and remain connected to the "real world" of the project managers.

My team

Well after all the compliments to my team in the last couple of chapters, I take some small credit myself in building such a team.

Now, to be fair, I inherited a small (three-person) team that was the original PMO (and I am more than delighted to say that they are still with me and form a major part of the Projects Team today). In fact, we have only added one more person to that team at the point of writing.

The Methods Team has been completely built up from scratch with a mixture of external and internal people led by one of my long-time (three companies so far) team members. Never underestimate the advantage of working with someone you know, trust, and who has proven themselves in the past. This team comprises four people.

And finally, the Outcomes Team is, again, led by another previous team member from a past company/PMO. This too is a completely new team with six members spread around the world as we needed team members regionally located for in-time/in-culture engagement with the local project managers and management team.

Overall, the full GPMO team has representatives in Canada, the United States, the United Kingdom, Germany, and Australia, a globale perspective being critical to our success.

DOI: 10.4324/9781003346470-11

Your team

You will build your PMO team in the way that you need to, of course, but the number-one lesson I have learned, from my own experience and from the experience shared with many other PMO leaders, is this: Get the best people you can.

In the Appendices, you can take a look at the "right" model I developed, and it talks to the PMO declaring that a PMO is all about having the "right" people and this means the very best people with most experience.

The most common failure in regards to the people in a PMO is to "do it on the cheap" and keep your best for other work, when in fact the very opposite should be true – take the best of the best and make them the heart of your PMO.

Think about it. Any PMO team member needs to be respected by anyone in the project community and their management, and they need to have the experience to assess, evaluate, support, challenge, review a project manager or programme manager, or offer mentoring and/or coaching. They need to contribute to recruitment, to career assessments, to deliver project health checks in a strong but supportive manner with a high degree of empathy. They need to think and lead at the forefront of project development and evolution. They should be team players, detail-oriented, and have excellent communication skills. They simply need to be the best.

Now that does not mean you can't bring less experienced people into the PMO team. You can. In my own, we have successfully worked with interns in the past, for example.

And the final lesson learned. Keep your people "fresh" and field relevant. You must not build an ivory tower PMO that declares and controls, but in reality, has lost touch with what really is going on at the project level and what is important today.

One approach to this is to regularly rotate people in and out of the PMO for new thinking, to help create "friends of the PMO", and to offer career growth - we use advocates and subject matter experts a lot.

Build a team

Here are some key areas to focus on:

- Hire the right people by looking for individuals who have the right skills, experience, and personality traits to work in a project management environment.
- Define roles and responsibilities but don't over architect them. It's better to let the team develop these themselves.

- Provide training and development to help team members develop their skills and knowledge further, which could include attending conferences, workshops, or obtaining relevant certifications.
- Create a positive team culture that fosters collaboration, communication, and continuous improvement, encouraging team members to share their ideas and insights, and celebrate successes together.
- Encourage team members to provide feedback on the PMO's processes and procedures regularly, which will help identify areas for improvement and ensure that the PMO is continuously evolving and improving.
- Support work-life balance[1] by providing flexible working arrangements, such as remote work options, or flexible work hours, which help team members maintain a healthy work-life balance and avoid burnout.

See more in Appendices – Team Building.

Note

1   In my own organisation, I am delighted to say that this is a very high priority for the company, and the team benefits from a very open and flexible working environment.

KISS and MURA

As an overriding principle, in all that we do, I have repeatedly shared (probably bored to death is a better way of saying it) with my PMO team that we need to "Keep It Simple". Even to the point of giving out notepads with this motto on it at meetings. KISS[1] is the most common acronym here.

In "Make Your Business Agile: A Roadmap for Transforming Your Management and Adapting" to the "New Normal",[2] I explored the sentiment of meeting complexity with simplicity.

> Leonardo da Vinci said, "Simplicity is the ultimate sophistication" and Shakespeare wrote "Brevity is the soul of wit".

> "Make everything as simple as possible, but not simpler" – is also attributed to Albert Einstein, so I am in good company with this attitude.

The underlying principle behind my first book "The Lazy Project Manager"[3] was working smarter and not harder and this can only be achieved through simple(r) application of thought, design, and action.

A recent Siegel+Gale[4] study revealed that employees in simplified work environments are 30% more likely to stay at their jobs because their time is spent on high-value work instead of endless meetings, reports, and emails.

Lisa Bodell,[5] in her book, "Why Simple Wins", offers up a definition of simplification (the MURA model) that involves the following four criteria:

1 M: Is it as minimal as possible?
2 U: Is it as understandable as possible?
3 R: Is it as repeatable as possible?
4 A: Is it as accessible as possible?

Try applying that to a process or two in your own PMO and organisation.

DOI: 10.4324/9781003346470-12

Most businesses start simple and (increasingly) then become complicated, but at the heart they are still simple; therefore, for me, it is entirely logical to aim to keep to the value of keeping things simple.

What sets great businesses apart from the rest is that there is that simplicity within, with a culture of KIS (let us drop the last "S", we don't need it – do we?).

Notes

1   KISS stands for "Keep It Simple, Stupid". The message is just as simple: Don't make your business processes any more complicated than they have to be. You'll just end up creating more work for yourself.
2   https://www.amazon.co.uk/Make-Your-Business-Agile-Transforming/dp/0367747081/ref=tmm_pap_swatch_0?_encoding=UTF8&qid=1684052008&sr=8-1.
3   http://thelazyprojectmanager.com/ The Lazy Project Manager, 2nd edition: How To Be Twice As Productive And Still Leave The Office (Infinite Ideas Publishing).
4   Siegel+Gale is a global branding company headquartered in NYC.
5   Lisa Bodell is the founder and CEO of Futurethink, a company that uses simple techniques to help organisations embrace change and increase their capability for innovation.

Chapter 2.8

Productive laziness

It seems right, at this point, to go back to my very first book, "The Lazy Project Manager",[1] for additional inspiration.

"Laziness - sloth: apathy and inactivity in the practice of virtue (personified as one of the deadly sins).

So lazy, or laziness, is mostly seen as a negative term, or at the very best, a term of selfish indulgence.

Productiveness, on the other hand, is seen as a very positive term. The ratio of work produced in a given period of time. Productivity relates to the person's ability to produce the standard amount or number of products, services or outcomes as described in a work description.

So, put the benefits of productiveness together with an intelligent application of laziness and you get 'Productive Laziness'.

Or to put it another way, you get the maximum output for any given input, with an eye to minimizing the input as well. Or, to put it yet another way, you get a lot of bang for your buck as some like to say!"

So, at all times, work smarter and not harder, and both "KISS" it and "MURA" it.

Best PMO Tip: The best PMOs will also consider working smarter and not harder as their core mantra.

Note

1   The Lazy Project Manager, 2nd edition: How to Be Twice As Productive and Still Leave the Office Early – Taylor, P – Publisher: Infinite Ideas Publishing 2009 and 2015.

DOI: 10.4324/9781003346470-13

Part 3

Building the image and the reputation

In Part 3, we will consider the marketing and the reputation of the PMO (whether that is the current reputation if you have an existing PMO or it could be the building up of a reputation for a new PMO, which is just as important as the business needs to "know" the PMO and understand what value it brings).

We then move on to ensuring that there is strong and effective communication and high engagement from the extended PMO "family".

DOI: 10.4324/9781003346470-14

Chapter 3.1

Branding the PMO and services

The value and purpose

Best PMO Tip: The best PMOs have a strong and recognisable identity within their own organisation.

One of the first things I did, much to the early confusion of the original PMO team that I inherited, was to engage with the marketing team and work with them to design a logo for us (based on the **Projects | Methods | Outcomes** naming convention that we had developed) (Figure 3.1.1).

Along with a logo (destined to go on anything and everything that the PMO published to our stakeholders) was a Word document template, a PowerPoint template, and an email signature.

My experience told me that the PMO that I was now leading needed a clear identity and we needed to remarket the revamped (new and improved) PMO to the rest of the organisation, and the best way to do this was to have a very clear and recognisable identity.

CERIDIAN GLOBAL PMO
Projects | Methods | Outcomes

Figure 3.1.1 Global PMO official logo.

DOI: 10.4324/9781003346470-15

Chapter 3.2

Marketing is your friend

In my book, "Project Branding: Using Marketing to Win the Hearts and Minds of Stakeholders",[1] I noted that I learned something very important a long time ago when I first started out in project management, and it was this: "*no matter how good a job you do, if you don't let people know, then most people just won't know.*

Then it seemed to come naturally to me to share what was going on within a project. With all the team members and to also share with others outside the project team. I quickly built up an affinity for and a relationship with the marketing departments within the organizations I worked for. In the early days this was mostly to make sure I gained access to all the goodies that marketing departments always seem to have on hand, the golf umbrellas, the pens, the polo shirts, etc. But that rapidly evolved into realizing how the marketing department, or at least the people with the marketing skills, could help me out on projects".

I further noted, "*Now surely, is also the time that you want to be the best. You want to be the best project manager you can be. And organizations want both the best project managers they can have and for these project managers to work in the most effective way possible and working effectively includes effective marketing of your projects*".

Now this can simply be extended to the world of the PMO. The identity, the understanding, the purpose of any PMO – especially if it's new to an organisation – requires some work from the PMO team and the PMO leader. The business generally does not or will not get it. They will not understand immediately what the purpose of the PMO is and therefore it's critical, in my personal belief, that you need to have support and help to promote to build the PMO to absorb it into your organisation's branding and the marketing department.

DOI: 10.4324/9781003346470-16

They get it. They understand it. This is what they do. This is their purpose.

So, definitely make friends with your marketing department.

Note

1   Project Branding: Using Marketing to Win the Hearts and Minds of Stakeholders' – Taylor, P – Publishers RMC Publications Inc. 2014.

Engagement and communication

Driving value to your organisation

As a starting point, try this short exercise.

- Consider what your PMO's reputation is amongst your PMO stakeholders
  - List the main stakeholders and then rate their perception of your PMO
  - Don't be afraid to talk to people – honesty is a good thing here
  - Surveys – Interviews – the more the better; do not make these internally subjective, the PMO's view; make these insights the reality of how people see the PMO today
- Now list the collateral and the communication channels that you have ready
  - Newsletters, promotional material on PMO/PM, success stories, etc.
  - Update sessions, knowledge-sharing "brown bag" sessions, attendance, and engagement at outside team meetings, etc.
- Finally, develop a simple plan of action on re-engaging the PMO stakeholders that don't have a high opinion of the PMO yet (and keep engaging with the ones that do have a good opinion, of course)
  - What collateral can you use and what do you need to develop to make this happen?
  - What means of communication are in place and what needs to change?

And if you feel you are in that "can't market/won't market" situation, just remember that you will almost certainly have some experts on hand … in your own marketing department. Reach out to them; I just know they will be happy to help you out.

DOI: 10.4324/9781003346470-17

Chapter 3.4

The "stop/start" principle

Stop first then start

One of the early challenges was to stop doing stuff that the PMO should not be doing (accumulation of all of those past "yes we can help" decisions made), along with historical "left-over" responsibilities, that other people probably didn't want to do.

Then it was about starting the things that the PMO should be doing, building out the roadmap for real.

All this took some time to shift from the "past" to the "future" and to re-guide the PMO towards the roadmap that was in place and acting as the guiding light.[1]

Now, if you are building a PMO from the ground up, then you won't have this issue as you will not be carrying history, but if you are rebuilding, then you will have expected work that may not fit in with your new PMO vision. Therefore, an early task is to a) identify this work and b) go find a nice new home for it somewhere else in your organisation.

Now I will freely admit that a) is a whole lot easier than b).

Then, step c) is to build a submission platform/process for people to submit work requests to the PMO with clear guidance that such requests will be assessed, prioritised, and may, or may not, be actioned; and even if they are actioned, they may, or may not, be "delivered" in the timeframe the requester expects.

It is all about control and management of your own destiny.

Start well and then keep going

Once the PMO "ship" has slowly changed course, then it is a matter of steering a smooth and true path through the organisational "icebergs" and business "storms", adjusting when needed, but always having in sight your own "North Star".[2]

DOI: 10.4324/9781003346470-18

This is, once again, easier said than achieved as there will be many demands on you from many people and many departments but, whilst you do your best to respond to the urgent asks, you must always keep in mind where you are leading with your PMO.

It is maintaining that vision and continually elevating your view of how things are going and anticipating what might be ahead whilst ensuring that the path is as clear as possible for you and your team (maybe continuing our ship analogy heading up to the crow's nest from time to time).

And having started this analogy, I will keep on going for just one more.

Franklin D. Roosevelt[3] said, *"a smooth sea never made a skilled sailor"*; in other words, if you have never experienced difficulties or challenges, then you aren't experienced enough.

One point on the whole "North Star" principle. Whilst you may have your PMO North Star, you cannot ignore the sponsoring organisation's own North Star and the two must be the same star and not a different constellation!

Notes

1   In fact, in honesty, there still is one area that we are working hard to pass away from PMO responsibility all this time later.
2   Metaphorically speaking, your North Star is your personal mission statement. It's a fixed destination that you can depend on in your life as the world changes around you. Polaris is a star in the northern circumpolar constellation of Ursa Minor. It is commonly called the North Star or Pole Star and is the brightest star in the constellation, readily visible to the naked eye at night.
3   Franklin Delano Roosevelt, commonly known as FDR, was an American statesman and political leader who served as the 32nd president of the United States from 1933 until his death in 1945.

Part 4

Delivering the road map

In this part, the road map deliverables are described in detail from the perspectives of the four focus areas that my own global PMO presents to all of our stakeholders:

- Project Community
- Project Methodology
- Project Academy
- Project Life

The community focus, the common framework, the project academy, project life cycle, project heath checks, and more are described: the design, the challenges, the evolution, and the impact.

Then we look to the future through an AI evaluation, before concluding with some insights into how our PMO had fun and how we engaged and communicated with each other.

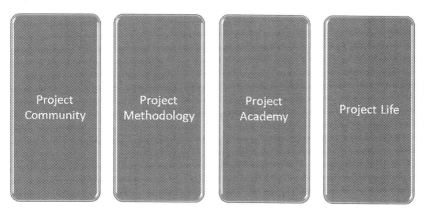

Figure 4.1 Global PMO four focus areas.

DOI: 10.4324/9781003346470-19

Project community

BEST PMO TIP: The best PMOs have the most experienced PMs in place and have a program underway to recruit the best PMs, to develop their existing PMs into the best, and to maintain this level of quality and experience.

Project community overview

[Peter: The following section on Project Community was written by Kathleen Rousseau from the GPMO who oversees this process for us and who does an amazing job]

In every great organisation, there's one contributing factor that makes the difference beyond all others, and that is the people – people making up the community. People working together with the same core goal move mountains, build bridges, and achieve great goals.

Just as people make up the Project Community, communication binds us together. That said, communication has been key to the Global PMO's success from day one. We work with and for the community, actively engaging in asking questions, listening, understanding, supporting the needs of the business, and producing strong results for continuous improvements.

The strategy to support our worldwide audience is to share project management–focused information with everyone at all levels in project-related roles in our services organisation, not just project managers. And why? This keeps the GPMO in tune with the business, making a difference, and supporting people's career growth and development.

DOI: 10.4324/9781003346470-20

Communication operations

Understanding history

Many businesses conduct all-hands meetings and the like. These types of meetings are valuable for disseminating business-specific messages to staff. Initially the GPMO hosted community of practice sessions, and periodic training sessions. While good in their own way, community of practice sessions, by their very nature, were narrowly focused, providing guidance on new and updated processes.

The GPMO made a significant difference by making a significant change. The GPMO evolved and expanded communications.

Developing a communications plan

When the GPMO first planned to evolve our communications, we discussed what our aim was in making a change. Essentially, we knew we wanted our audience to keep their finger on the pulse of the business; hence, the establishment of "pulse" sessions. We wanted to provide inspiration from outside the business within the project management industry, which became our "inspire" sessions. And we knew that periodically we'd want to provide an opportunity to bring the global community together in "summit" sessions to share strategy, thought leadership, and knowledge.

Our goals further developed into a plan to make our communications readily accessible to our worldwide audience, and to make it easy for speakers and content contributors to participate in bringing their message to the market. And a challenge we wanted to overcome led to developing a platform to promote the sharing of small bundles of information in concise and engaging ways.

Knowing your audience: everyone's your audience

Before we discuss specific types of sessions, it's important to understand the audience.

Being the GPMO means that our purview is projects. The things we do naturally revolve around projects and project managers, but projects are not managed in a vacuum. They are supported and must be understood by more than just project managers. It was with that mindset that the GPMO determined that everyone is our audience and communications are project management focused.

Defining session types

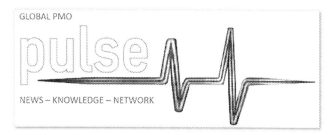

Figure 4.2 GPMO pulse session logo.

Pulse sessions

Businesses will always have a need to share news, updates, and things of general interest. People are used to getting small bundles of information effectively presented in the news and social media, so pulse sessions are designed to provide:

- Structured community engagements and knowledge sharing
- Supportive learning experience, primarily PM focused
- Open access to all
- Frequency: Monthly
- Duration: 60 minutes with two to three separate topics
- Accessibility: Live presentations, recorded and published separately as digital video files for ease of consumption 24×7 (podcast-style)
- Outcome: Improved learning experience and sense of community

Figure 4.3 GPMO inspire logo.

Inspire sessions

Created as a completely new concept for the business, the GPMO developed the inspire series to bring in some of the project management industry's leading authors and keynote speakers from around the world to share their inspiring thoughts with our audience.

- "Keynote" style speaker engagements to provide thought leadership
- Scope: Project management–related topics from different points of view
- Open to all
- Frequency: Monthly or bi-monthly
- Duration: 60 minutes, one speaker
- Accessibility: Live presentations, recorded and published for 24×7 access
- Outcome: Broadened insights and thinking around critical topics

Inspire sessions have been so successful in providing stimulating stories about the speaker's project management experiences and observations, that the GPMO further expanded this offering to include customer inspire sessions. In this series, we have identified dynamic customers with interesting implementation stories, willing to join their counterpart on our business side to discuss not just those things that went well, but also those crunchy conversations they were able to get through to overcome challenges along the way for a successful outcome. The customer inspire session format really resonated with our audience at all levels, and feedback has included thanks from the field for the open, honest, and transparent communications.

You can see a list of some of our great external speakers in the appendices.

Figure 4.4 GPMO summit logo.

Summit sessions

Another new concept to support the business is the summit. This communication format varies in frequency and duration depending upon business needs. Successful summits have been held annually or bi-annually to align with significant business updates and rollouts. They have been scheduled across three to five days, depending on topics to be covered. An inspire session with an external keynote speaker has been included in each of the GPMO's summits. Whenever possible, repeat sessions have been conducted live in various time zones, and recorded for the larger audience to ensure that everyone has access to the same message.

- Rich and engaging sessions to showcase strategy, thought leaders, and knowledge
- Scope: To bring the global services community together to share the message
- Open to all
- Frequency: Annual or bi-annual, multi-time zone
- Duration: Multiple sessions with varying durations
- Accessibility: Live presentations, recorded and published for 24×7 access
- Outcome: Improved transparency and sense of GPMO community

You can read all about a very special summit, called 'The Event', in Part 6.

Building a project community program

Can this go from concept to live in the snap of a finger? No, this is a lot to unwrap, so instead, consider what you can do first. Start there and build up the program.

Do you have an existing series of meetings that could be refreshed? Consider replacing the series with pulse sessions.

- Review upcoming topics and presenters. Do you think these topics could be broken into small, consumable presentations of 15–20 minutes, with 5–10 minutes for questions and answers? If the answer is yes, and you can bundle a couple topics together, and then you have the makings of a pulse session.
- Meet with your guest speakers in advance to promote the new format and gain their buy-in. Advance planning sessions are likely already a part of your process and will continue to be needed for pulse and other new sessions.

Planning – producing – polishing – publishing – promoting

Project Community has a lot of moving parts, and having a leader dedicated to the process is important for managing the program – coordinating topics and speakers, planning, developing, publishing, and promoting.

Planning – keeping things organised

If your business is like ours, you have an abundance of topics to choose from and speakers that are happy to support and contribute their subject matter expertise. How you convert this to actual results is through careful planning.

A leader's role is to identify pertinent and timely topics and contributors, to guide them through expectations, and to demonstrate patience and flexibility while working with dynamic, busy contributors to develop and deliver their message.

Create a project plan using your preferred application and format, and make it as simple or detailed as you like, as long as it works for you, and:

- Log all planning meetings, including:
  - Weekly or bi-weekly program-level communication planning sessions
  - Contributor-specific planning sessions (including a tech session/dry run session, to validate readiness, review tech features of your teleconference application, and to do a brief dry run)
  - Include brief notes about your meetings within the log
- Log all pulse, inspire, and summit sessions, and consider color-coding to differentiate session types
- Have contributors provide their presentation deck by a specific date/time, as a backup in case of technical issues the day of The Event, and document when this is due within your log
- Schedule time to finalise the deck, adding in the contributor's slides, and tweaking your talk track

Approval for engaging external speakers

Another critical consideration is that if you plan to include inspire sessions in your Project Community program with external guest speakers, be sure to research first to determine if this is allowed by your business. Confirm if your organisation has or needs to create a speaking engagement agreement for your use. And be sure to confirm whether or not your organisation will allow payments to external speakers. Answers to these questions are critical before beginning any other steps in the inspire session planning process.

Producing – creating a great audience experience

How you prepare for and deliver communications to your audience is a big deal. There's a whole entertainment industry built around creating a great audience experience. Clearly, communications we're discussing here are on a much smaller scale by comparison, but the audience experience is still important.

How can you schedule and present sessions that are impactful for your audience? It all starts with three key steps:

1 Schedule sessions well in advance and ensure your meeting sessions have everyone who joins the call muted.
2 Be sure to open the line at least 30 minutes early in all cases, to run through your last-minute tech check for both you and your guest speakers. Doing this ensures you have time to confirm any embedded video and audio files are working perfectly and that audio levels are good.
3 Consider logging into the session from a second computer for visibility from a meeting participant perspective. This helps you reduce stress and build confidence.

These steps lay a great foundation and provide a reliable and repeatable process you may leverage for all types of sessions.

Scheduling sessions

Pulse sessions should be held regularly, at whatever frequency makes sense for you and your organisation. Here, the GPMO schedules them the second Thursday of each month and varies when needed for holiday and business reasons. Recurring invitations are sent out to the audience to secure time on their schedules. As the session date nears, the invitation for that session is updated to include the session topic, and to ensure that we're capturing everyone that's joined the organisation since the invitations were first sent out.

Inspire sessions and customer inspire sessions are planned for alternating months and are only scheduled after speaking engagements are confirmed with our keynote and guest speakers. Note that the GPMO schedules the sessions in whatever time zone is convenient for the speakers. This means employees around the world will have some opportunities to join GPMO-hosted events live, rather than relying exclusively on videos.

Summit and other sessions are scheduled, and invitations sent when announced.

In addition to sending invitations, the GPMO leverages its SharePoint[1] site to include an events calendar that is maintained to reflect all scheduled events for the year.

CREATING THE BASIC PRESENTATION TEMPLATE

The GPMO has created a PowerPoint presentation deck template that is just a "shell" into which content is added. The template consists of six slides, as described below. Preparing this type of template in advance allows you to include key elements and your basic talking points, which you'll alter from presentation to presentation.

- Slide 1: This is leveraged "pre-session" to indicate the meeting will begin shortly
  - You can start simple with just a slide that may include music
  - The GPMO has created a video loop with music that's used to good advantage
    - While letting early joiners know the meeting will begin shortly, the GPMO's video also promotes other things happening in the business, updates on key initiatives, introduces PM new hires, and celebrates key accomplishments
- Slide 2: This is a "host welcome" slide that's used to share a brief message before the session and recording begins, welcoming attendees, asking everyone to remain on mute during the call, and to drop any questions in the chat
- Slide 3: The session's title slide should be updated to include the appropriate session logo (pulse, inspire, or summit). Here, the host begins recording, triggers music to play for 20–30 seconds, welcomes everyone to the call, and announces the topic
- Slide 4: The introduction slide should include photos, names, and titles for The Event's speakers. As the host, you'll make the introductions and may reference their bio, if suitable, before handing things off to the speaker for their presentation
- [Speaker's Content]: Plan to add the speaker's slides between slides 4 and 5 in the template
  - Note: Typically, the speaker then shares their screen to present the deck and drives through the presentation, or the host does so on the speaker's behalf, depending on what was pre-arranged
- Slide 5: This slide may be labelled for questions and answers or may simply include your company logo. Depending on the circumstances, as the host, you'll thank your speaker for sharing their message, thank your audience for joining or viewing the video and supporting the GPMO, and let them know where and when the video will be published
- Slide 6: Use your final slide to trigger music to play for 20–30 seconds before stopping your recording and ending the session

REFINING THE TEMPLATE FOR THE SESSION TYPE

Preparing for all types of sessions, it's good to start with the basic presentation template.

- For inspire or summit sessions, use the basic template and include the appropriate logo on the title slide
- For the pulse deck:
  - Use the basic presentation template
  - Use the appropriate logo on the title slide
  - Copy and paste to combine two or three templates together into one presentation deck
  - Since you'll only need the "pre-session" and "host welcome" slides at the very start, delete any other "pre-session" and "host welcome" slides from the deck
  - Remember that this is flexible, and you can make this fit your needs for your pulse sessions, and it's easier to have this template prepared in advance and you can delete any section that's not needed

RECORDING OPTIONS

While we're talking about presenting and producing, consider that not all communication sessions need to be live sessions. Sometimes, to provide timely communications to the organisation, it may be best to conduct a closed recording session by working directly with your contributor to create the video.

See Appendices for some more technical guides around options for this.

Our GPMO team has successfully produced pre-recorded sessions in several instances and indicated to the audience as it was published that the session was pre-recorded.

POLISHING – DELIVERING PROFESSIONAL RESULTS

There are many options for teleconference calls and webinars, and many applications for completing after-session production refinements. The information provided reflects how our GPMO manages the process, and we acknowledge that this is not the only way to do so.

TOOLS/APPLICATIONS

- Microsoft PowerPoint is our preferred format for presentation decks, and for the simplicity in embedding audio and video files.

- Unsplash.com is our preferred application for capturing copyright-free images
- Zoom meetings are our preference vs. webinars
  - Following meetings, we recommend uploading the mp4 video file rather than just sharing a Zoom link and password with the audience
    - Why? Because some communication sessions may be leveraged long term as training content for the organisation, and mp4 video files are accessible long term, whereas Zoom links have a shorter shelf life[2]
- Camtasia is the application we use for production work on our Zoom mp4s. Files are easy to import and export, and the application includes several options to use multiple tracks, trim as needed, and refine the video using audio and visual effects, transitions, animations, captions, and even add voice narration, if needed. Completing production results in polished and refined mp4 video files that we publish for our organisation.

PROCESS

- Pulse sessions: As we record two or three topics in each of our pulse sessions, we've developed a process for the best viewer experience:
  - Our live session recording runs continuously
  - Instead of sharing announcements and such, after each topic, we review them only at the end of the final topic for our live audience
  - Then, during production, as the video is split into separate pulse topics, we add the announcements at the end of each video
  - This takes a bit longer in production, but it helps us ensure that everyone viewing videos will receive the same announcements, regardless of if they opt to view all or only select videos
- There are many options, but we use Camtasia for this process due to its simple process for trimming videos, when needed, and to use audio features to fade sound in and out, and visual transition features to fade video out to a black screen at the end of our recordings.

Publishing – making content accessible

The GPMO publishes pulse, inspire, and summit session videos internally using the PM community in Yammer/Viva Engage,[3] which is connected to the GPMO SharePoint site. Members of the PM community receive notifications when videos are published. The full list of invitees in the organisation is also sent a monthly email that includes links to all GPMO-recorded sessions to date for the current year.

Promoting – building your brand

Project Community, when used in the manner outlined here, not only strongly supports the business, but it also demonstrates the GPMO's engagement, alignment, and comprehension of the business. When leaders think of how best to communicate with the field, they think of the GPMO. This journey the GPMO embarked upon to evolve our communications has done more than that. It also had a positive impact on evolving our business.

Figure 4.5 GPMO Community Events.

Just some examples of our success:

- Community Success – 40,000+ all channels
- Success of the Pulse sessions – Live attendees are 200 regularly, but we have had up to 400+
- Success of the Inspire sessions – 200+ live attendees with some of the best PM speakers in the world
- Growth of Inspire sessions to now include our client stories

Beware of communication overload

The phrase "information overload" was first used by Alvin Toffler[4] in 1970, when he predicted that the rapidly increasing amounts of information being produced would eventually cause people problems. His suggestion was that although computer processing and memory is increasing all the time, we – the humans – that take on board and process and use that information are not getting any faster in relation. We have, effectively, reached our limits, as was discussed in my book "The Social Project Manager".[5]

A such, the PMO team has carefully assessed the overall communication load to our global project manager community, with a view to making sure that no one person is impacted through "information overload".

I noted in another book, "Real Project Management":[6] "Uncontrolled communication is as bad as no communication at all; therefore, all of your communication should be planned and considered."

We continue to do this. Even at the point of writing, we are investigating a new format for information flow through the community that we feel might well be more effective and easier to consume.

Notes

1 SharePoint is a web-based collaborative platform that integrates natively with Microsoft 365 (previously, Microsoft Office).
2 Camtasia is a software suite, created and published by TechSmith, for creating and recording video tutorials and presentations via screencast, or via a direct recording plug-in to Microsoft PowerPoint.
3 Yammer is an enterprise social networking service that is part of the Microsoft 365 family of products – Viva Engage is a new app, integrated in Teams, that surfaces existing and new employee experiences powered by Yammer services.
4 Alvin Toffler was an American writer, futurist, and businessman known for his works discussing modern technologies, including the digital revolution and the communication revolution, with emphasis on their effects on cultures worldwide. He is regarded as one of the world's outstanding futurists.
5 The Social Project Manager: Balancing Collaboration with Centralised Control in a Project Driven World – Taylor, P – Publisher Routledge 2015.
6 Real Project Management: The Skills and Capabilities You Will Need for Successful Project Delivery – Taylor, P – Publisher Kogan 2014.

Chapter 4.2

Project framework

BEST PMO TIP: The best PMOs are the custodians of a dynamic framework of method to assist PMs in the delivery of projects. This includes not only process but also templates and guidance and knowledge sharing.

[Peter: The following section on Project Framework was written by Kim Dimauro and based on her work with her Methods team – Erika, Lauren, and Sally]

Why even have one

Implementing a methodology in project management is essential for several reasons.

It provides structure, consistency, and a systematic approach to project delivery, ensuring that projects are executed efficiently and effectively. It is a critical guide to inexperienced project managers, and it is a reference point for experienced project managers.

Without a methodology, projects can become chaotic, disorganised, and prone to failure.

Let's explore some of the reasons why projects need a methodology: Standardisation and consistency:

- A methodology brings order and consistency to project execution, enabling teams to work together effectively towards common goals.
- By adopting a methodology, project teams have a standardised set of processes, tools, and techniques to follow. This consistency allows for better collaboration, seamless communication, and a shared understanding of project objectives. As a result, stakeholders can have confidence in the project's progress and outcomes.

DOI: 10.4324/9781003346470-21

Scalability and adaptability:

- A well-defined methodology accommodates projects of varying sizes, complexities, and regions, ensuring adaptability to meet diverse project requirements.
- A methodology should be flexible enough to scale according to the project's needs. It should account for different project sizes, types, and geographic locations. This adaptability ensures that project teams can tailor their approach to suit specific client goals and expectations, whether it involves a simple lift and shift or a complete business transformation.

Best practices and industry standards:

- Adopting industry best practices within a methodology improves project governance, mitigates risks, and enhances overall project success.
- Incorporating project management best practices and industry standards into a methodology is crucial. It ensures that projects are executed using proven approaches, reducing the likelihood of errors, and increasing the chances of success. By aligning with established practices, project teams can leverage the collective wisdom of the industry to drive better outcomes.

Feedback-driven improvement:

- A methodology should be a living framework that evolves based on feedback, continuously improving, and addressing the changing needs of projects.
- Feedback plays a crucial role in refining and enhancing a project methodology. By actively seeking input from project teams and stakeholders, organisations can identify areas for improvement and incorporate valuable suggestions. This iterative process ensures that the methodology remains relevant, effective, and aligned with evolving project management practices.

Change management and adoption:

- Change advocates are vital for successful methodology adoption, as they promote and support the transition, driving organisational buy-in and engagement.
- Change management is a critical aspect of implementing a new methodology. Change advocates, recognised as experts in the field, play a key role in driving adoption. They provide support, communicate

benefits, and champion the methodology, enabling smooth transitions and fostering a positive organisational culture around the new approach.

To summarise, a well-defined project methodology is necessary for effective project management. It provides structure, consistency, and scalability while incorporating best practices and industry standards. Feedback-driven improvement and change management ensure that the methodology evolves, while the development of a support model facilitates successful roll-out and adoption.

Remember, as Margaret Mead[1] once said, "Never doubt that a small group of thoughtful, committed citizens can change the world; indeed, it's the only thing that ever has." The successful implementation of a project methodology relies on the commitment and collaboration of all stake-holders involved.

> We make life difficult, and then we try to solve it. My methodology is to simplify things and share them with life examples. Ashish Vidyarthi[2]

Know what you have and know what you need

This brings us to the start of our methodology journey some 18 months ago.

One challenge we faced was a variability in methodology use across the world. There was the "official" method, but regions had, through necessity, created deviations to address their tactical needs.

The first step was to understand the current methodology status and determine the challenges with it – the intention was not to invent from scratch but to identify what worked and was good (file under "keep") and what was not working (file under "replace" or "improve") and what was missing (file under "create").

And so, we began by conducting a survey of the services/implementation community. The community consisted of project managers, consultants, sponsors, and more (for your own world add in other groups as relevant).

Based on these survey results, and further analysis, they informed us that within our methodology, the "one-size-fits-all" approach wasn't working, and we needed to take a more global approach and holistic approach, and account for a variety of project sizes, types, and regions in our project world.

It is identified that the methodology challenge was the number-one issue with project managers around the world.

We now had a foundation of a plan:

- Align our global delivery approach through a flexible, three-stage project framework

- Create a consistent experience that scales with project size, complexity, and risk
- Emphasise project management best practices
- Simplify project checkpoints with predefined success criteria
- Define a predictable release cycle to support change management over time

Build on the vision

Based on all this insight, we then created a vision to establish the new methodology, ensuring to simplify the delivery experience by introducing a flexible, delivery framework. This framework should create a consistent experience that allows scale based on the complexity of the client and the client's requirements.

For example, some of our customers are looking to lift and shift, while other customers are looking to transform their business by adopting a full solution. By understanding the customer's unique goals, you can utilise the methodology to create the right fit and experience for that customer. Do this by taking a modular approach to key areas of delivery, allowing the delivery team to select the right tools at the right time for their project.

When you design and build, emphasise project management best practices by applying industry standards to how you govern and deliver projects. Gather best-in-breed artifacts, templates, and tools to provide that global delivery experience.

And start "selling" the vision throughout your organisation in order to connect, convince, and gain cooperation.

Make it happen

Next, establishing a strategy to create the new methodology was key, and we completed this by mapping out the current process. Once this was done, we formed several different working groups, with key stakeholders who would be involved throughout the renovation process, to ensure regional acceptance. Following this approach provided a path to creating the necessary training (more on this later).

Now, you may have spotted a critical word in the above, "renovation".

Here, we are back to marketing and taking people on a journey with you. If we had presented this as an all-new methodology, that would have sent the message that everything in the past was worthless – and it most certainly wasn't. By talking "renovation", we made sure that the good was seen as of value, and retained, and we presented an opportunity to improve and to make it more modern, relevant, and forward looking.

Waterfall or Agile

OK this is going to be short.

We are hybrid. Best of all worlds. End of story.

What is interesting for the (near) future is the work we are doing on automating elements of the process/method in order to digitally accelerate our projects.

How was it made?

The framework was developed on a dynamic SharePoint site. This was tailor-made using components and custom web parts, allowing for different project paths. Think of a shopping cart experience. As the user selects products to implement, the framework renders back the appropriate methodology based on the selection. This is also dynamic by region, and complexity of the project.

We think that this is pretty neat.

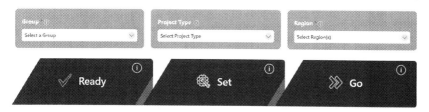

Figure 4.6 Project Methodology Structure.

And we created a simple-to-understand structure.

Built on the experience developed from our history of thousands of successful deployments, the simple, but effective, "Ready, Set, Go" approach optimises project success.

1   The Ready stage is where we connect with the customer and share the project path to success, with a clear starting point and practical tools to begin the great work.
2   The Set stage drives overall project success through an iterative process of requirements gathering, solution building, and testing. We've laid the foundation and now it is about ensuring the solution exceeds the customer's expectations.
3   The Go stage covers go-live activities through to project closure. It's where all the amazing work done throughout the Ready and Set stages comes to fruition. By using our methodology, it provides a consistent experience that scales with project size, complexity, and risk.

A template library is also necessary for content management. This allows the user to easily filter documents by region, stage, and document name and also creates a clear version control for when documents are updated.

Also, pretty neat.

Project Tools

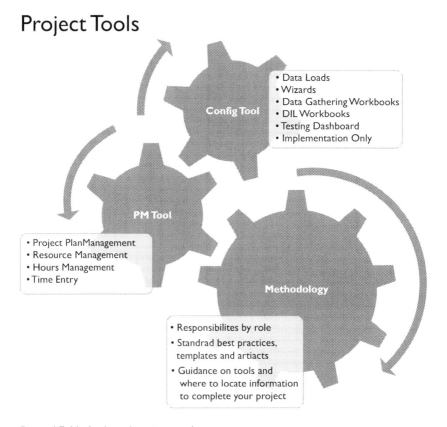

Figure 4.7 Methods and project tools.

Prove it before sharing

Once the new framework was established, it was crucial to vet it both internally and externally before any live release.

For our renovation, we had four live projects as part of the pilot, and we created a virtual "road test" to further validate and test.

From the pilots, the feedback was positive from customers and the project teams alike and thanks to their feedback we have been able to make the methodology even better.

On the virtual "road test", we had 60 internal participants representing all regions and roles from project managers and sponsors to delivery managers and consultants. Almost 40% of the feedback from those drivers came from North America: 36% from APJ[3] and 24% from EMEA.[4] It was a fantastic cross-section of our business user base and we received more than 200 pieces of valuable feedback. We were able to take immediate action on at least 75% of that feedback prior to launch and continue to incorporate changes for the next methodology release (we talk about release plan later on).

We worked with our Training Team for six months to establish training objectives for each role. During our sessions, training needs analysis, a purpose statement, and KPIs[5] were determined, and in turn, they created a self-paced training course to support the methodology at launch and for all (future) new project managers.

Change management is key

Change management was achieved by using advocates across regions and teams to provide an extra layer of support. The role of a change advocate in an organisation is someone who promotes and supports change and champions their efforts toward making the change happen.

Recruitment of advocates is a key component to the adoption of the new methodology (check out the "Adoption" chapter). These change agents are generally considered experts in the field and are therefore trusted proponents and supporters.

Providing advocates clear expectations, i.e., attend team meetings, familiarise themselves with the methodology, and how feedback from the field will be handled, is critical.

And such, advocates can ensure that there is a clear path for the team that created the new methodology.

The final stage is to develop a support model for the roll-out of the new methodology.

1   Participants need to complete the training courses created.
2   Advise all users of established office hours to answer questions and get support (in their own time zone ideally).
3   Set up a site where it's easy for individuals to collaborate and get answers to their questions – and share knowledge as well.
4   Communicate, communicate, communicate.

We also created a feedback form, which allows individuals to share with us their ideas for improvement of the methodology, as we know this is not a one-time-only creation but a living framework.

All such feedback feeds into the road map, and once vetted, will be incorporated into future methodology releases.

But key in the immediate aftermath of a release like this is the effectiveness of the hyper-care put in place, so all new users are well supported in a time-critical manner and that there are no obstacles to adoption put in place.

It should be noted that (and you will hear more about this in the "Outcomes Project Life" chapter) our project health checks were adjusted to take into account the new version of the methodology and the adoption/understanding levels will be measured through these activities as another channel of help, and understanding.

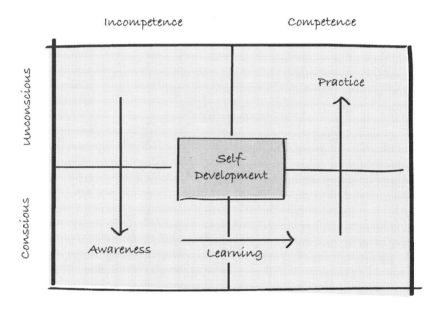

Figure 4.8 Methods self development.

Our aim is for all project managers, and other project key players, to become what is known as "unconsciously competent" – the best at their art. But we also have to cater for the full journey of personal development for everyone from the "newbies" to the career aspirational.

The above chart was referenced in Peter's book "The Lazy Project Manager" and is a variation of the Johari Window[6] of understanding and capability.

This approach is further supported by the Project Academy, which you will read about in the next chapter.

Marketing

As we have already seen, marketing is a friend of the PMO and we connected with the team to develop a range of collateral, including a high-quality, high-impact, promotional video (well, two actually, one for our internal audience and one for our external audience).

What did we say? Well, we said:

Our process is collaborative, focused, and trusted.

• Collaborative – Our team will partner with our customer's project team to work towards a successful go-live.
• Focused – When our clients are busy with other business priorities, our team will be there to keep them on track.
• And trusted – Our process incorporates best practices and world-class service with a trusted and proven method.

And we concluded:

• Success is measured by how quickly and effectively customers can go live with our technology.
• Our methodology for projects provides a standardised path for delivering a positive go-live experience by ensuring alignment with the customer's goals throughout the project life cycle.

The videos were great and simply presented to people the power of the offering.

Share and celebrate

You will see in Part 6 the manner in which we launched the renovated methodology and the impact that it has had.

But suffice it to say that it is critical to share and celebrate such major events in a suitable way and you should consider for your own PMO deliverables the best and most appropriate way of doing this.

And programs

Everything that we have presented so far has been project focused, but earlier this year it became apparent that we also needed a methodology to guide our program leaders as business grew and we began many more enterprise-level, multi-country programs (or programmes, depending on where you are reading this).

As such, in an incredibly short timeframe of just four months, the team took a working group and crafted a program delivery framework, which was new since this was no renovation but a start-up initiative.

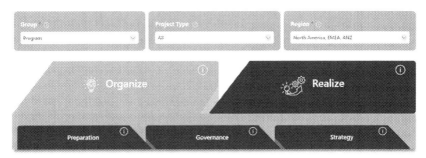

Figure 4.9 Program methodology "Organize".

Our program framework, designed to work in association with our project framework, has two major phases – "Organize", which is all about the preparation for a program launch, the governance set-up, and the program strategy.

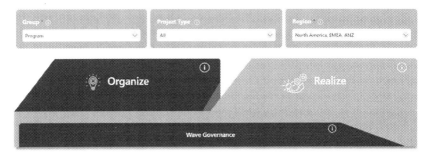

Figure 4.10 Program methodology "Realize".

And then the "Realize" phase which oversees what we refer to as the "waves" of deployment – the projects themselves.

Our program methodology delivers everything that our project-level framework delivered, and in fact integrating with this project methodology, and delivering to the same quality level it was also part of the launch event.

Amazing work from a truly awesome team.

The future road map

Finally, as we noted, to support the project managers (and program managers) and offer up truly relevant best practice, then the methodology will be structured like a product with quarterly releases made up of both "fixes" and improvements, together with real enhancements.

And we publish a road map to showcase what is coming in the future for clear visibility to all our stakeholders, and this goes at least 12 months out so a clear path to the future.

The future is, we truly believe, bright!

Notes

1 Margaret Mead was an American cultural anthropologist who featured frequently as an author and speaker in the mass media during the 1960s and the 1970s.
2 Ashish Vidyarthi is an Indian actor who predominantly works in Hindi, Telugu, Tamil, Kannada, Malayalam, English, Odia, Marathi, and Bengali films.
3 Asia-Pacific, Japan.
4 Europe, Middle East, Africa.
5 Key Performance Indicators.
6 The Johari Window is a framework for understanding conscious and unconscious bias that can help increase self-awareness and our understanding of others. It is the creation of two psychologists, Joseph Luft and Harrington Ingham, who named the model by combining their first names.

Project academy

BEST PMO TIP: The best PMOs sponsor training and facilitate communities of practice to promote PM best practices in their organisations. Such communities of practice provide PMs with a forum to share their knowledge and share experiences and have a common, open, progressive career path opportunity.

[Peter: The following section on Project Academy was written by Bill Snow based on his work with his Projects team – Kathy, Brandy, and Ahmad]

Project Academy overview

The Project Academy strikes the balance between individual and organisational learning goals. It is intended to support our people in an environment that offers structured learning experience, aligned with individual learning needs of the community members.

The Global PMO is an influential leader in the business and an integral player in the execution of the organisational strategy. It provides an edge for the creation and deployment of a structured educational program through the Project Academy, which allows the collective business to prioritise education, engage internal and external expertise to capture and document processes and best practices, and recognise and celebrate learning.

Education and learning are key to growth and evolution – without the information, guidance, knowledge, and insights, it is hard for anyone to find a path to be the best they can be.

The Project Academy provides training capabilities that allow people to identify learning plans that are suitable for their growth. However, the creation and deployment of a Project Academy can not only be viewed as a mechanical entity that delivers training; rather, it must be thought of as a strategic transformation initiative to promote personal development and

DOI: 10.4324/9781003346470-22

career growth in the organisational priorities. This higher prioritisation around education and knowledge will deliver ROI through higher engagement, career satisfaction, reduced employee turnover, and better execution for better client satisfaction and business results.

Establishing major ventures such as Project Academy requires a purposeful approach to define the objectives, strategy, and design tools for successful execution and benefits realisation. This is a key strategic pillar to improve the value to your employees and drive a deeper level of engagement.

The famous Richard Branson[1] quote that he penned in 2014: *"Train people well enough so they can leave, treat them well enough, so they don't want to"*. The statement from the founder of Virgin Atlantic carries a lot of meaning, and it should undoubtedly be the driving force of every entrepreneur.

The origins of the quote probably go back to Henry Ford,[2] who reportedly said, *"The only thing worse than training your employees and having them leave is not training them and having them stay!"* The sentiment clearly strikes a chord – we need to invest in people, even if it increases the risk of them leaving.

The community concept

Communities in organisations are an essential part of business success. Healthy and engaging communities reflect an elevated level of alignment to organisational objectives and values. They also provide a set of metrics for measuring the level of adoption of processes and their effectiveness.

At the core of communities in organisations are the employees. They are the main drivers of any process or practice, and consequently business success. Organisations today are more sensitive to the employee well-being and are increasingly investing to enhance the employee experience and satisfaction. The employee engagement model is shifting towards more remote work and flexibility, which requires organisations to look for new innovative methods to interact with their people and keep them engaged.

Sustaining active communities in business will promote a culture of collaboration, learning, and growth that will contribute to employee well-being and ultimately business success.

The Project Academy model

As organisations shift to leverage the power of active communities to enhance employees experience and potential, it is important to embed cultural and behavioural values that prompt learning and engagement. Global PMO can play an important role in achieving the active communities in organisations by offering a scalable and flexible model to allow project management community to organically communicate, learn, and innovate. This model is designed to support people experience through the following:

• Learning, growth, and development
• Collaboration and knowledge sharing
• Feedback and engagement
• Social promotion
• Sense of belonging

We call this model the "Project Academy", designed specifically to support our organisation's strategy, attract new talents, create a learning culture, and activate more coaching.

Find your "why"

The first step is to define the "why".

You must understand the business needs and gaps and determine where the Global PMO can fulfil and bridge these gaps through an Academy model. The "why" should be crisp and clear to develop relevant solutions by The Project Academy to leverage the community and people power.

Once the "why" is defined, we then create a vision and mission for the Project Academy that resonates with the "why" and that will matter to our organisation and our people. The vision and mission must be aligned to organisational strategy and cascade from Global PMO objectives. They must be inspirational and leveraged to gain the buy-in of key stakeholders.

The vision and mission statements must simple; however, they should be specific, achievable, and relatable to evoke excitement and maximise user support and influence to improve business and individual results success.

Makes Work Life Better

MISSION

Creating a structured environment
for education, growth, and standards

VISION

Enabling people to help
organizations transform and create
exceptional HCM experiences

Figure 4.11 Project Academy Mission and Vision.

Target Operating Model

After you have created the vision, it is important that shortly after, or in line, with the creation of the vision, that you determine how you will turn vision into action and real value. This comes in the form of the "Target Operating Model" and the overall program strategy.

Creating a viable operating model is critical to govern how the Project Academy will deliver services and value to the organisation. The Project Academy is not a one-off initiative; it is an ongoing "entity" that provides set of services to specific users to achieve specifics goals. Thus, it is important to have a structure in place to manage the Project Academy operations, continuous enhancements, and innovations to better service its "customers".

The operating model lays out the structure and actionable road map to achieve the Project Academy vision and mission. Consider the following when identifying the operating model for your own business:

- Formulate the vision and mission
- Develop the key goals and objectives of the Project Academy
- Define the key principles that become the "beacon" that guides us through how to operate and ensure achieving objectives
- Define the pillars or focus areas of the Project Academy
- Define the initiatives associated to achieving the objectives and create their charters
- Define key performance indicators (KPIs) and associated targets to monitor the Project Academy performance
- Define the Project Academy model: what services will be provided and how
- Define the value chain of Project Academy services
- Develop the governance model, interaction model, and associated roles and responsibilities

Every organisation is unique and has its own challenges and needs. The Project Academy target model should be flexible and cater for the specific business requirements. It is up to the organisation to develop the strategy for the Project Academy that fulfils their own needs; however, this strategy should always be aligned with overall objectives and values of the organisation and also support the career path enablement and employee experience.

Below are examples of four pillars or focus areas that might be applicable to any organisation:

1 New hires: Reimagine the new hire and onboarding experience to welcome our new talents in an effective and innovative way to make them ready to deliver and excel. Prepare them for success by creating a

sense of belonging, planning for personal growth and fulfilment, availing all required tools and resources, and creating the support structure to easily sail off in a long journey within the organisation.

2 Tenured Employees: Support existing employees and community members to structurally expand knowledge, develop new skills, and successfully navigate through the career path.

3 Learner Validation: The learning experience must be relatable, genuine, and authentic. Learning must make an impact and learners should be confident they can practice the knowledge they acquire.

4 Sustainable/Interactive Community: A key objective is to create an interactive community that can build the Project Academy brand, effectively contribute to overall well-being, in continuous proximity with learners and its ecosystem, and innovate and feed in new ideas. The Project Academy is about people; it is important to have community members engaged to interact, evaluate, and provide feedback; share new ideas; develop relevant content; and enhance overall learning experience across the organisation.

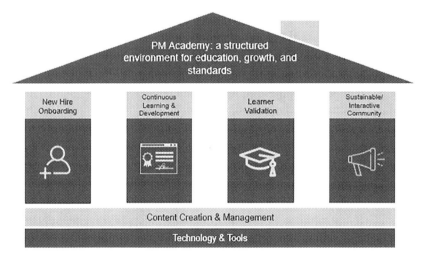

Figure 4.12 Project Academy Foundations.

The four pillars above should be supported by foundations that will ensure:

- Innovative and seamless processes for content creation and management that will ensure leaners receive relatable and reliable material.
- Technology and tools to offer the community easy mechanisms to consume the knowledge pool.

Project Academy principles

We discussed earlier that is key to define the key principles as part of defining the target operating model. These principles will govern the shape and standards of the Project Academy products and services.

Consider the following as an example list of principles. The Academy will:

- Synergise outstanding learning experience and badge offerings through partnering with HR and internal training capabilities and service providers
- Utilise leading learning tools and concepts
- Tailor the learning experience to the needs of different career levels in order to support continuous learning and individual career progression
- Adopt the following key design concepts in developing and delivering learning topics, content, and qualification badges:
  - Consumable, Engaging, Meaningful
  - Theoretical + Real-World Skills
  - Flexible Learning + Enable Career Growth

These principles represent the Project Academy's overarching guidelines and promise to the end users and community members.

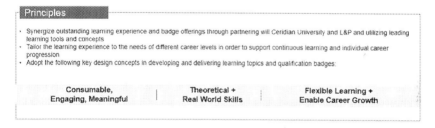

Figure 4.13 Project Academy Principles.

Designed with your success in mind

This is the motto of the Project Academy: "Designed with your success in mind".

It is about people, their experience, and their success.

The Project Academy must deliver real value that matters to the end user, specifically the project management community. The overall design must resonate with them and provide a great user experience. Define the personas that reflect the different user categories and related experience. This becomes the centre of who the Project Academy is catering and creating value to. For example, consider the following three personas to start:

- New Joiner – The value proposition is rooted in three key areas 1) recruit talent by showcasing an exceptional onboarding experience through the Project Academy; 2) create a guided, rich, and consistent onboarding experience; and 3) drive a high level of engagement sentiment.
- Career Growth – Most project managers aim to grow and achieve high levels of title and status in their job family and even move to another job family. The value proposition for this group is that the Project Academy can help enable this growth, including better understanding of the business, performance, etc. By creating a badge model whereby, the badges are aligned to the various promotional opportunities, project managers will have a clear path to learning that will support their desire to get promoted
- Continuous Learner – This is the casual learner who is interested in expanding knowledge on different topics. The Project Academy will provide a variety of offerings and courses that can fulfil the persona's learning needs

More personas can be identified. You will need to assess your own organisation's learning needs and activities and develop offerings and services accordingly.

Success is not vertical. Project managers cannot sustain success by just growing their technical knowledge only; they want to expand their skills in different aspects and areas to help them better make a difference and maintain growth and success. Accordingly, the Project Academy must

Figure 4.14 Project Academy Balanced Learning.

provide balanced learning experiences that cover different verticals such as process and methods, people management, business acumen, technical acumen, leadership, and others that can be applicable to the organisation's training needs.

Building, maintenance, and operations

Most PMOs are not designed to have an educational organisation built natively within their operations. A PMO is generally not built to be a training organisation – they don't usually have certified educators, professional instructional designers, professionals that design certification programs, nor the tools and technology. As such, organisations should not underestimate maintenance and operations activities required for managing the Project Academy business as usual, such as building content, maintaining content, managing users' profiles, data gathering and analysis, user journey, issuing badges, and others.

Therefore, a key part of strategy is to create an ecosystem of partners and experts, as well as identify and even purchase technology that will support great, quality outputs and experiences. The PMO should not try to be an expert at everything – this is a time to outsource and leverage any economy of scale!

An ecosystem could include:

- Internal partnerships with organisations such as internally focused training teams, customer facing training teams, instructional designers, human resources, marketing, subject matter experts, etc.
- Evaluate current available technology, such as learning management systems (LMSs), Microsoft tools (e.g., SharePoint, Teams, etc.), video creation and editing tools, and, certainly, exploring the world of artificial intelligence.
- Assess operational needs, such as reporting, publishing, maintaining content, data gathering and analytics, etc. – and then determine if there are others in the business that are already doing these things and if you might be able to outsource to them for economy of scale.
- External partnerships may need to be purchased; things such as learning content that you want to buy, subscriptions for content providers, web designers to help build websites, marketing experts to build marketing assets, digital badge distribution services, etc.

Building the academy assets and tools

In the early start up stages, you can expect to spend a good amount of time and effort creating your systems and collateral. This includes:

- Designing and creating websites
- Creating a logo and marketing collateral and mood boards for templates (PPTs, celebration templates, etc.)
- Establishing KPIs and reports for data gathering and analytics
- Curating and creating content

Curating and creating content

One of the main values of the Academy is the learning content that you serve up to your audience. This can come in many forms, including things such as white papers, YouTube videos, short and consumable video clips, formal self-paced training, formal instructor-led training, etc. Generally, you may want to consider fourtypes of learning paths:

- Short, consumable content
- Single courses
- Learning plans made up of multiple forms of course work (e.g., short content, courses, case studies, etc.)
- Certifications, such as proctored exams

Your PMO can expect to spend most of their time working through the process to determine what the priority topics are and how best to deliver them. Depending on the personas that you are focused on and your objectives for the Academy, you may choose different paths. An example may be:

- Career Growth Persona – you may choose to create learning plans that are rich and deep in content that require an investment of time to become an expert in a given topic – these plans would be aligned to a career growth path (e.g., promotion). This investment in achieving a learning plan may be rewarded and recognised through a digital badge or some other formal means
- Continuous Learner Persona – this may be someone that wants to gain knowledge in specific topics – this type of learner may only need the short consumable content and/or single courses that provide targeted knowledge with less time investment

In order to deliver these, your strategy should likely include an approach to both curate and create content. We believe the best training strikes a balance between theory/concepts and practical learning. Theory and concepts are important to convey thought-provoking ways of thinking or approaching a topic and the practical learning gives the learner real-world insights that they can apply immediately.

As you create content, one of the biggest investments and challenges will be partnering with SMEs (subject matter experts).

Driving usage, adoption, and change

The start-up of the Academy is a change initiative.

You can invest a lot of time in building the Academy and all of the great content, but if the audience does not see the value or prioritise the time to take training, then your program will surely suffer or fail. The launch must include fanfare – it should be exciting – you will want to create marketing videos, create your theme and brand, create your logo – when people think of Academy or see your logo, what do you want them to think? How would they describe Academy? How do you want them to feel?

People must feel like they are part of something – part of a club. They need to believe they will get something they cannot get anywhere else and something that they feel is for them and will make them better.

A critical approach to drive adoption and buy-in is to identify change agents and influencers in the field. These are leaders and peers in the field that can champion the messaging and influence others to believe in the academy and invest in the Academy.

The user experience must also be frictionless and easy. The process and journey for that learner needs to be well-thought through, which is a component of your operating model strategy, and translate into a great learning experience. In addition, the Academy must be easy to work with and supportive of the experience. You may want to consider a "concierge service", where a learner can engage the PMO to get a level of individual and white-glove service.

You should also consider mandating training as part of your strategy to drive engagement and usage – if you want to be a "directive" or an "authoritative" GPMO, mandating critical training is one lever to pull. For example, budget management is critical to effective project and program management, so building and mandating obtaining a badge for budget management is one way to drive usage, education, and business alignment around critical topics.

Attracting and impressing talent

The Academy is a valuable and unique asset; not every company invests in a program like this. In labour markets where companies are doing all they can to attract and retain key talent, the Academy is one of the appealing things to leverage to draw in talent.

As part of the Academy, you likely want to have a new hire onboarding program, designed to not only provide purposeful training, but give that new project manager a sense of community and belonging from day one. Many seasoned professionals and project managers have experienced a lacklustre initiation[3] and onboarding period, so imagine reading a job description that showcases a company that has a new-hire Academy that intentionally sets someone up for success!

A new-hire onboarding program should have elements of formal training with a variety of training formats (reading, listening, doing), case studies, mock exercises, shadowing, and observation periods.

A well-structured and engaging new-hire program should give the new hire space to learn and grow, build confidence, and gain an immediate sense of community.

You should also consider a strong element of celebration and recognition through the form of graduation ceremonies and digital badging.[4]

Marketing

As we have already seen, and will continue to see and say, marketing is a friend of the PMO and the Projects Team and is also connected with our marketing team to development a range of collateral, including a high-quality, high-impact, promotional video (well, two actually, one for our internal audience and one for our external audience – just like the Methods Team did).

An exciting platform helps our people learn and grow in a vibrant and proactive community by using effective content management and technology.

The Project Academy vision is to empower people to help organisations transform and create exceptional HCM experiences.

And our tagline was *"Designed with your success in mind"*.

Social marketing

As part of the sharing celebration of achievements when people achieve badge-level certification, we used Credly[5] to create and share digital badges of achievement.

And we encourage sharing on social media of these digital badges by individuals to doubly raise their profiles and to promote our own organisation. As an exmaple our very first badge release saw over 400 people complete the asociated course and achieve a badge – a significnat number of whom posted their celebration on social media. A real win-win situation.

Ongoing value

Academy cannot be a "one and done".

There must be an ongoing road map and ongoing investment in a road map. There must be a program around 1) building and bringing new value and 2) marketing and communication to restated and reinforce the value. It is a constant showcase of value to the organisation and an effort to keep them excited, engaged, and keep everyone contributing the community of learning.

One longer term of the Academy may be to find a way to monetise some offerings. Most PMOs are a cost centre that do not generate revenue – they bring plenty of value, just not in the form of revenue. One way to change that is to find a way to monetise the Academy to become a revenue-generating PMO (or at least get closer to cost neutral). There may be training and learning services that might be appealing to audiences such as your customers or partners. Take, for example, if you have a unique methodology to implementing your software/services – a training certification or badge may be something that your customer might invest in for their team prior to launching the implementation of the software/services. See more about monetising a PMO in the appendices.

Notes

1   Sir Richard Charles Nicholas Branson is a British business magnate and commercial astronaut. In the 1970s, he founded the Virgin Group, which today controls more than 400 companies in various fields.
2   Henry Ford was an American industrialist and business magnate. He was the founder of Ford Motor Company, and chief developer of the assembly line technique of mass production.
3   You have probably had mixed experiences yourself in your career.
4   Brilliant for social promotion and recognition/celebration.
5   https://info.credly.com/ Credly is an end-to-end solution for creating, issuing, and managing digital credentials.

Project life

BEST PMO TIP: The best PMOs have consistent, repeatable PM practices across the enterprise. All projects are held to the same standards and requirements for success. They have also eliminated redundant, bureaucratic PM practices that have slowed down projects.

[Peter: The following section on Project Life was written by Thomas Neumeier based on his work with his Outcomes team – Matt, Jost, Mitch, Anand, and Jason]

The beginning

Where should we begin – maybe in the past?

Looking at the PMOs in the world, they usually start to define themselves in one of three main directions: tactical, strategic, and organizational. We have all probably gone this path as well and always found something missing or broken. It doesn't fit with the balanced PMO approach, but actually it's an easy way to measure success; and if you haven't built your own PMO yet, it's still a good enough starting point.

So, what are we? Why is our PMO different?

Well, as you have read something about the Projects and Methods Team, they are clearly strategic and centralized, building the next level of our future for the methodology, training, and tools. Looking at the Outcomes Team, as part of a centralized Global PMO, we think globally but act locally, to support our project managers to drive successful project outcomes. And to do so – depending on the need – we act either tactically, strategically, or organizationally. We are based in the same region and in

DOI: 10.4324/9781003346470-23

the same time zone, we speak the same language, and we understand the culture. Most important of all, we all have real experience, and through this we are bring with us a lot of experience and a heavy toolbox to support our work.

We found our way through defining three main key focus areas under which we always pin new initiatives:

- Connection and Relationship: We believe the basis for a strong project management community is building connections and growing relationships through ongoing communication. To do so, the Outcomes Team drives initiatives by connecting and gathering insights regarding day-to-day issues that affect project managers, so that we can continue building initiatives to improve and enhance Project Life within our organization.
- PM Life Cycle: This key focus area is the overall arching process map for any step of project management in the organization. It has plugged in initiatives and targeted communication points based on the phases below:
  - Pre-Start Phase
  - Start Phase
  - Performing Phase
  - Close-Out Phase
- Project Health Check: It is based on the premise that project success in any organization may improve through the uncomplicated process of asking a few questions periodically of the project manager and the project team to assess the health of a project.

So, back to the three main directions (tactical, strategic, and organizational) – we are doing all three based on lessons learned from previous PMOs we ran.

And also, as a lesson learned, we are really local in the region our project managers are located. Typically, the biggest mistake companies and PMOs are making is "to stay" in the base of home country – operating remotely. they end up only supporting the project manager community in their own time zone and not that of the actual project managers. Getting on early morning or late evening calls is not something helpful for project managers outside of this "master" time zone, especially in this fast-paced and multi-project environment. Project managers need a support structure in place within their local time zone.

Just think this through. If project managers are running multiple projects simultaneously (very common in our world at least), their normal workday is already packed. So we cannot assume that they are keen to jump out of the bed hours before they usually start or stay up late and impact their private lives, just to attend an information call or discuss any project questions or issues.

Well, you might immediately think, "let's do recordings" – am I right?

The correct answer here is "no", but to be fair, this lesson we had to learn ourselves as well, and it was painful, especially for those expected to watch them. Indeed, there was a running joke at this time: "Great, yet another recording". And, of course, the PMO was not the only group using this record-and-watch-later approach.

So, in fact, the project managers ended up with lots of emails every week with multiple recordings to watch. All important, all mandatory, and all supposedly very helpful, but in reality, all too much, too superficial, and most critically, excluding the ability to be interactive in any way.

So, what should you do if your business is truly global?

If the decision for going global is made, investments to support those regions need to follow immediately, especially the support for those resources bringing in the expected revenue.

We established regional offices of the PMO as part of our overall GPMO structure. It was required to stream down any information and updates, all to be shared in the time zone of the local project managers and to offer them "in-time" support. No magic, just a simple communication layer to support and protect at same time, which was very well received by every member of our project community.

Looking at our current Outcomes Team, we are six people represented in three regions. We came up with a ratio of approximately 40 PMs to one regional PMO Outcomes Team member, considering holiday and sickness coverage, as the optimum level.

Connection and relationships with the field

Building both (especially in a virtual environment) isn't easy, but absolutely necessary. You need to know your project managers, what they think and what they are struggling with, as well as what is going well and how they approach things.

To clarify, this connection and relationship needs to exist between the project managers as well.

While building a house, you must have a strong foundation. In our case, we established two communication channels for each project manager, regardless of their segments.

Email – our PMO support email address can be used for all types of questions and queries at any time. Well, you might think now, "What's different to what you have?" – nothing really from a technology standpoint, but as the Outcomes Team is based all over the world, we can provide a 24/6-ish support with an SLA[1] of less than eight hours (and in fact our response time is less than three hours).

But, our PMO Teams chats are more important[2] for NA/LATAM, EMEA, and APJ. It's where the magic happens and we share updates, monitor, and sometimes answer questions. Most is done by the project managers themselves, as they support each other by sharing information and answering questions.

The last point for the foundation is our initiative, "GPMO 1:1 with Everyone". Actually, it's a little bit of a misnomer as actually we spend 45 minutes with three project managers and one Outcomes Team member in each session, and we undertake this twice a year. It is certainly time-consuming once you include the post-work and evaluation of feedback, but we are currently looking at a participant rate of 84% – far above what you could get with a survey and as we connect with every project manager, we will, for certain, build up a crucial relationship.

Following our conceptual house, looking above the foundation, our rooms are separated by regions and segments. We have eight monthly PMO outcomes calls with project managers to share updates and the latest news, what's coming next in the GPMO and project manager world, and provide a platform to ask specific questions or for project managers to present topics. These calls are accompanied by eight PMO outcome calls with managers of the project managers and another three calls with senior leadership. Depending on need or request, we host deep dive sessions on any specific topic. Those calls are in addition to our GPMO pulse and inspire sessions (see "Community" chapter for information) with a total effort for each project manager of less than three hours per month.

Finally, looking at the roof of our "house", we conduct one survey, "Voice of the PM", per year to collect anonymous feedback on the work our PMO does.

Summarizing these initiatives under the key focus area, we are very well organized as a result.

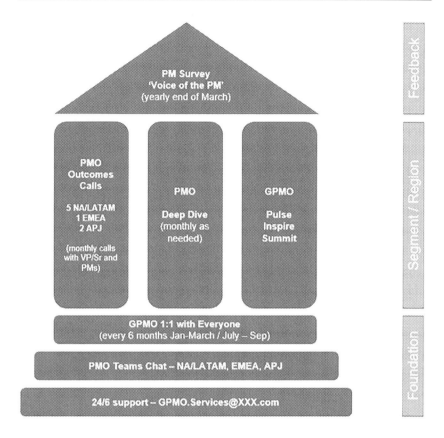

Figure 4.15 House of connection and relationships.

PM life cycle – from start to finish

Similar to a project life cycle, which contains the complete duration of a project from even pre-project tasks to the final one, we have designed multiple initiatives around the life of a project manager in our organization, which starts even before the project manager starts to work for us.

We have split this key focus area into four phases, aligning different initiatives to each:

- Pre-Start Phase
- Start Phase
- Performing Phase
- Close-Out Phase

So let us start with the first phase.

The Pre-Start Phase contains all processes and artefacts prior to a candidate signing an employment contract. From generating and opening a job requisition, including a predefined standard job description, through to officially becoming part of our large worldwide community and network; pre-selection of possible candidates to go through the interview process up to conducting interviews to ensure their project skills meet the expected level. The exit criteria for this phase are a signed contract and our PMO welcome email to the new hire.

The Start Phase covers all onboarding activities and courses for the newly hired project manager, including courses around the organization and product, the Project Academy Flight School training (see previous chapter) specifically designed for project management knowledge including our tools and methodology, and a ten-week shadowing program as part of the field readiness with a peer coach assignment. This phase is finished after a health check is undertaken, typically three months after a new project manager's first project assignment, to ensure our new project manager is set up for success.

Let's skip the Performing Phase for now, as this is the biggest one, and concentrate on the Close-Out Phase first. Unfortunately, we have also an attrition rate in our project management community (just being honest here – but it is significantly reducing we are pleased to say), which we are monitoring carefully by region and project manager level. Actually, this is one of our important KPIs, as costs for new hires and the loss of experience can be critical to the business.

Some project managers are getting promoted into other positions or into their well-deserved retirement, and others may leave the organization. Regardless of the reason, we are reaching out to them to get their direct – non-HR related – feedback to understand and improve. Finally, we support the delivery management during project transitions to ensure these are following our process standards.

Ok, so let us focus on the Performing Phase now. As noted, this is the biggest phase, which includes a lot of different initiatives and the connection points to our other two main focus areas: Connection and Relation and Health Checks.

During a year in the Performing Phase, we are potentially connecting with each individual project manager 20 times:

- Twice a year through the "GPMO 1:1 with Everyone",
- Once a year through the PM survey "Voice of the PM"
- Twice a year through the Health Checks
- Up to 12 times through the PMO outcome calls
- At least five times through PMO deep dive sessions
- And an additional 12 times through the GPMO pulse and inspire sessions

That's a healthy number of times where we interact with project managers directly, and not to forget all the interaction via our regional PMO Teams chats. Through this, we have a good and solid overview on how each individual is performing from an experience and competency perspective, understanding what help they might need or how they can share their knowledge with others – it works both ways.

In addition to this, our KPI dashboard provides more insights on each project manager and in aggregation into each delivery group. You will read more about this KPI dashboard later in this book, but for now, all this information ties into our project manager scorecard that we are using for suggestions around upskilling or even promotions. Specifically for promotions, we have designed a dedicated assessment program to ensure transparency and equal treatment wherever the project manager operates in the world.

So how would you categorize this focus area? Correct, it's a mixture of being tactical, strategical, and organizational.

Health checks and "in-flight" corrections

> BEST PMO TIP: The best PMOs ensure that quality assurance actually delivers quality.

[Peter]

Audit or Health check or Post-Mortem? I covered this in another of my books 'Get Fit with The Lazy Project Manager',[3] which was all about project health and the many aspects of delivering and measuring the health of any project.

Now there is often a lot of confusion with regards to the three, incorrectly interchangeably used terms – Audit, Health check, and Post-mortem.

For clarity and intelligent application, here are the dictionary definitions:

- Health check: A thorough physical examination; includes a variety of tests according to need.
- Audit: An official examination and verification of accounts and records, especially of financial accounts, a report, or statement reflecting an audit; a final statement of account or the inspection or examination of a building or other facility to evaluate or improve its appropriateness, safety, efficiency, or the like.
- Post-mortem: Occurring or done after death, of or relating to a medical examination of a dead body (also autopsy). Discussion of an event after it has occurred.

For further clarity and greater intelligent application, think of it like this:

• When was the last time a person survived multiple autopsies?

Another way to look at the difference is this. Consider the "humanity" aspect:

• Health check – all about the person; therefore, very high on the humanity scale
• Audit – far more about the process or the facts, numbers, and inanimate objects; therefore, very low on the humanity scale
• Post-mortem – high on the "human" side of the equation admittedly, but all a little too late in the day I would suggest and so extremely low on the humanity scale

So let us please discard post-mortem once and for all and focus on "Audit" versus "Health check".

OK, well, projects are about people so what do the people think?

One company I worked with used to do audits but re-branded them as health checks simply because people were nervous about an 'audit'. It seemed to all be about checklists and compliance and seeking out deviant behaviour, and as a result there was low acceptance/high resistance of such audits. Health checks on the other hand were actually considered warm and friendly and a whole lot less scary. They were seen as supportive and guiding and a positive thing – rather than the negativity of an audit – and as a result there was high acceptance/low resistance to these occurring on projects, especially by the project managers.

It is up to you, but I go for health check every time, and health check it is in my PMO.

[Back to Thomas]

So now as we have found the right name for those assessments, let us focus on the reason for conducting them.

Looking at our projects – especially the small and medium-sized ones – neither PMI[4] nor IPMA[5] would consider these as projects at first glance. Our project managers tend to drive those projects via a task list with the same approach over and over again. The unique point on these are the different circumstances of complexity and client focus. In our multi-project landscape, our project managers have to deal with highly complex environments and at the same time with different levels of project management maturity on the client side. In other words, our project managers need to be flexible enough to support projects with evolved project management experience from the client side through to highly complex projects with an "accidental" project manager without project management experience on the client side. That makes all of our projects unique and requires a sensitive way to approach the health check.

Although we are using a generic report to assess the project, each health check is done individually and most importantly at the same level. There might be good reasons that a project manager did not follow our methodology or did not use our provisioned tools as expected. As long as core project management principles are applied and all information documented and stored in our project repository, we don't rate this project low. Quite the reverse – we are looking more into the details to understand and see if we could use this approach as best practice to evolve our current standards. And this is the reason why automated project health checks don't work well.

[Peter]

I did assess an automated tool some time ago and the main issue with it was that it successfully identified areas for improvement or current weakness but failed to record the positives and, as real people, project managers deserve to be told what was good as well as what could be improved.

[Back to Thomas]

We have multiple trigger points for initiating a health check:

- Request by senior leadership or by a project manager themselves
- Random selection by the PMO to ensure two health checks are undertaken with each project manager each year
- Automated trigger points backed into the methodology, mainly raised due to client feedback
- Automated trigger points as part of the project manager onboarding process with a health check three months after the first assignment
- Automated trigger points through schedule and cost variances

But to make it clear, this main key focus area is built upon a very tactical approach.

But, and this is a very critical question linked very much to the earlier chapter of the right PMO team, who should perform or conduct these health checks?

As mentioned before, it is very important to do this at the same level, whoever is delivering or overseeing such health checks. That means the person (and never ever call them the project health checker) must have on the one hand plenty of experience in project management (respect is crucial when offering such feedback), especially in a services environment. Plus, this person must be equipped with excellent soft skills to lead, guide, and coach the project manager during this exercise. Again, at the same level means we are not finger pointing, blaming, or policing the project manager during a project health check but supporting, guiding, and aiding them.

Finding the right resources is hard and, through experience, our preferred strategy is to use only existing internal resources who have a minimum of three years in project delivery.

But, please don't underestimate external knowledge. It's important to get new, fresh thoughts in to a PMO to widen your perspectives and drive improvements; otherwise, the risk would be to end up in a deadlock caught in outdated processes.

Finally, those resources need to be local and in the same time zone as the project manager, as well as speaking the same language. If you are running a global business and start performing these health checks with resources based in one single region, you will miss the regional and cultural differences. It would immediately be recognized as corporate supervision – the Project Police[6] – a situation you need to avoid.

This is a journey, and we are already developing our project health check version 3.0 – there is always room for improvement.

Notes

1   A service-level agreement (SLA) sets the expectations between the service provider and the customer and describes the products or services to be delivered, the single point of contact for end-user problems, and the metrics by which the effectiveness of the process is monitored and approved.
2   Microsoft Teams is a proprietary business communication platform developed by Microsoft, as part of the Microsoft 365 family of products. Teams primarily competes with the similar service Slack, offering workspace chat and videoconferencing, file storage, and application integration.
3   Get Fit with The Lazy Project Manager: How to make sure your project is as healthy as possible and does not become the "ex-project" of tomorrow – Taylor, P – Publisher TLPM Publishing 2014.
4   The Project Management Institute is a U.S.-based not-for-profit professional organization for project management.
5   IPMA is a federation of about 70 Member Associations (MAs) who develop project management competences in their geographic areas of influence, interacting with thousands of practitioners and developing relationships with corporations, government agencies, universities, and colleges, as well as training organizations and consulting companies.
6   Peter – I have repeatedly said in my various books and presentations that a good PMO should not act or be seen as the dreaded project police.

Chapter 4.5

Exploring the future with AI

> BEST PMO TIP: The best PMOs are not stuck in their ways, but are proactively exploring opportunities to be better and to serve their sponsoring organisation and their project community better.

In my recent book "AI and the Project Manager: How the rise of Artificial Intelligence will change Project Management", I explored this topic to some degree, but I think many, if not most of us, have been shaken by the incredible rise in the AI "possible" in business and life. ChatGPT[1] being just one example that most of us will have heard about, if not played around with (Microsoft and Google also moving fast in these areas). Appendices has the ChatGPT summation of the author's views on PMOs – just for fun naturally.

The change coming was further raised in my own Gartner article, "Artificial Intelligence—Threat or Aid to Project Managers?", I noted "Artificial intelligence (AI) has opened up a torrent of projects and possibilities, helping organizations reimagine how we live, work and play. And it's happening at a rapid clip. According to a Gartner survey, * 80% of executives think that automation can be applied to any decision. But what about the impact on project delivery and project management?

It seems unlikely to be minimal: Gartner says that 80% of project management tasks will be eliminated by 2030 as artificial intelligence takes over.[2]

Let that sink in for just a few moments. If you are a project leader, 80% of what you do today you won't be doing in just seven years' time (if not earlier). What will you be doing instead? Will it still be in project management?"

I offer up some practical advice on engaging and using AI in a second article, "How AI Can Help Project Managers Bring the Focus Back to

DOI: 10.4324/9781003346470-24

People", https://www.capterra.com/resources/focus-on-people-with-ai-in-project-management/.

But back to the challenge and the change.

In the *Harvard Business Review* article "How AI Will Transform Project Management" by Antonio Nieto-Rodriguez and Ricardo Viana Vargas,[3] they noted, "This new generation of tools will not only change the technology on how we manage projects but will change completely our work in the project. Project managers must be prepared to coach and train their teams to adapt to this transition. They should increase their focus on human interactions while identifying technology skill deficits in their people early and work to address them. In addition to focusing on project deliverables they should focus on creating high performing teams in which members receive what is needed to allow them to perform at their best."

So, with regard to my PMO, we certainly did not want to get behind the curve on this and so we commenced a series of evaluations of AI technology to see if it a) could work for us and b) could deliver significant performance (and financial) benefits.

Mitch[4] from my team is helping lead these investigations and thoughts for us, and he offers up this summarisation:

As a company, we must always innovate in a pragmatic way; otherwise, we will lose to future competitors ....

Benefit

- It's a tool that can help enable decisions more efficiently, but it is only a tool; it can be used correctly or incorrectly.
- It bridges the gap between those with and without knowledge.
- Helps sort through mass volume data sets to interpret the information.
- Natural language queries are intuitive and are a future direction.
- Enables efficiency by removing menial tasks (project admins eliminated).
- Raises the bar for delivery of content; using AI tools should help, but some may not have the skill set to leverage it or resistance to change.

Risk: Trust

- Are we conditioning people to trust an anonymous AI system over people and management?
- Someone coded the AI; to make inferences based on conditions ... What bias was coded into the AI?
- Acceptance – do we accept the recommendations unconditionally?
- Unanticipated results – AI may lead to information or conclusions that are incorrect; some may not be aware of this.

And moving on to a non-human "expert", ChatGPT[5] generated this response when asked "Why should companies not adopt AI in the project management world".

While AI can offer numerous benefits in various industries, including project management, there are several reasons why companies might be hesitant to adopt AI in this particular field. Here are some potential concerns:

- *Complexity and Implementation Challenges: Incorporating AI into project management processes can be complex and require significant investments in terms of technology, infrastructure, and skilled personnel. Implementing AI solutions may involve integrating with existing systems, training employees, and addressing compatibility issues, which can be time-consuming and costly.*
- *Lack of Human Judgment and Intuition: Project management often requires human judgment, intuition, and decision-making skills that may be difficult for AI systems to replicate accurately. While AI can automate certain tasks and provide data-driven insights, it may struggle to handle nuanced situations, adapt to unforeseen circumstances, or evaluate complex human interactions effectively.*
- *Reliability and Trustworthiness: AI systems rely on data, algorithms, and models to make decisions. If these components are flawed, biased, or based on inaccurate or incomplete information, the AI system's output may be unreliable. Ensuring the reliability and trustworthiness of AI systems in project management becomes crucial, as incorrect decisions could lead to costly project failures or delays.*
- *Lack of Transparency and Explainability: AI algorithms can be complex, making it challenging to understand how decisions are reached. This lack of transparency and explainability can be a concern, particularly in project management, where stakeholders need to understand the reasoning behind decisions to gain trust and make informed choices. Regulatory requirements and ethical considerations may also demand explainability, particularly in sensitive projects.*
- *Organisational Resistance and Cultural Shifts: Introducing AI into project management requires a cultural shift within the organisation. Employees may be resistant to change or feel threatened by the idea of AI taking over their roles or responsibilities. Building trust, providing adequate training and support, and fostering a positive mindset towards AI can be significant challenges for companies.*
- *Data Privacy and Security: AI systems in project management often rely on collecting and analyzing vast amounts of data. Ensuring data privacy, security, and compliance with regulations like GDPR (General Data Protection Regulation) and other industry-specific requirements becomes crucial. Companies need to carefully manage*

and protect sensitive project-related data throughout the AI implementation process.

- *Unforeseen Consequences and Errors: AI systems are only as good as the data they are trained on. If the training data is biased or flawed, the AI system can produce biased or inaccurate results, potentially leading to unintended consequences. Additionally, AI systems may be susceptible to adversarial attacks or unexpected errors that could impact project management processes and outcomes.*

It's important to note that these concerns can be mitigated or addressed through careful planning, transparency, continuous monitoring, and appropriate use of AI technologies. The decision to adopt AI in project management should be based on a thorough evaluation of the specific needs, risks, and benefits for each organisation.

My personal view is that this is inevitable; there is no fighting it; there is no resistance to this; it will come to pass; and we will have to change, adopt, or go find something else to do with our lives. My advice is to get ahead of AI and project management curve if at all possible, or at least stay up with it as best you can.

See also 'The State of AI in PM: Leading Project Manager Unpacks Brand New Capterra Data' for a real insight in to the state of AI in project management today.

https://www.capterra.com/resources/data-on-ai-in-project-management/

I remain positive that it will be a huge benefit to the profession of project management.

This is an exciting future for project managers – independent thinking – creative – compassionate – collaborative – and even more successful than today, powered by AI and focused on people.

Notes

1 ChatGPT is a natural language processing tool driven by AI technology that allows you to have human-like conversations and much more with the chatbot. The language model can answer questions and assist you with tasks, such as composing emails, essays, and code. An example of ChatGPT output can be found in the appendices (just for fun).
2 Gartner Says 80% of Today's Project Management Tasks Will Be Eliminated by 2030 as Artificial Intelligence Takes Over, Gartner https://www.gartner.com/en/newsroom/press-releases/2019-03-20-gartner-says-80-percent-of-today-s-project-management#:~:text=By%202030%2C%2080%20percent%20of,%2C%20according%20to%20Gartner%2C%20Inc.
3 https://hbr.org/2023/02/how-ai-will-transform-project-management.
4 Mitch Rajmoolie (see appendices).
5 https://chat.openai.com/ created on 19th June 2023.

Chapter 4.6

Team building (fun)

BEST PMO TIP: The best PMOs have a whole lot of fun as part of their work, all of which contributes to a positive mindset and a good work/life balance.

Richard Branson of Virgin Group said, "Have fun, success will follow. If you aren't having fun, you are doing it wrong. If you feel like getting up in the morning to work on your business is a chore, then it's time to try something else. If you are having a good time, there is a far greater chance a positive, innovative atmosphere will be nurtured ... A smile and a joke can go a long way, so be quick to see the lighter side of life".

And fun is definitely out there in the project management world, as I noted well in "The Project Manager Who Smiled".[1] Having the right sort of fun project environment can be good for you as well.[2] "Done right you will have set the acceptable parameters for fun in your project, both in content and in extent, and you will have engendered that spirit amongst your project team to the point where, one day, when you are the one on a low, they will come up and make you smile."

We, as a team, have used many opportunities to have fun, take a break away for the work side of life, and generally enjoyed ourselves.

Just a few ideas that you might consider:

Obviously, since we had a logo (#marketingisourfriend), we needed some T-shirts and this was the amazing design that our marketing team came up with, in association with our supplier:

DOI: 10.4324/9781003346470-25

Figure 4.16 The official team logo.

We ran a Christmas/Holiday Season quiz (used Menti[3] for this one).
We had a hosted virtual scavenger hunt.

One team had a team building make-your-own-pizza evening.

We ran the marshmallow challenge (one of the best team activities ever – see appendices for a guide on how to run this yourself).

In one of our most-of-the-team-present meetings,[4] which ran over a few days, each country was represented by "owning the day" and so we had food and "goodies" from Australia, chocolates, and a quiz (Germany v UK), and we had a full-on BBQ feast and celebration from the USA, as well as a host of goodies from Canada (loved the Maple Syrup team). Share those cultural differences!

One team member ran a personality-type exercise, which was both fun and insightful.

We joined in a virtual exercise journey as a collective (and we all got medals) https://www.theconqueror.events/lejog/.

We even planned a fun music video recording with fancy dress (popstars versus rappers), but sadly, we got "hit" by COVID on the team and had no choice but to abandon this activity (saving that one for a future date).

We made some really great videos (part of the day job to be fair, but we still had fun putting them together).

And we ran a "project managers theme tune" competition. Stage 1, all project managers could suggest great songs (think "Always look in the Bright side of Life" or "Mission Impossible" or "Under Pressure" and you'll get the idea). Stage 2, we compiled a list and the project managers all voted for their top three favourites. Stage 3, we ran a special pulse session with a top-ten countdown, chart-style, playing snippets of each song.

Partly inspired by a couple of things, we found https://youtu.be/sWIG44FaCww ('The Project Management Song' by Conrad Askland) and https://youtu.be/Ej66TiINaRc ('The Project Manager Blues' by Frank Saladis) and Project Manager: The Movie https://youtu.be/-3-oWaFIoV8?si=MJmM0xqdx6fMt8nI

We have had some amazing team meals in some amazing places.[5]

Plus, love this one, we have our own cocktail (created by a professional mixologist and recipe shared on a team call).

I give you the "GPMO" (see what we did there?).

The GPMO

Ingredients

Gin 35ml
Pineapple Juice 25ml
Mure (Blackberry Liqueur) 15ml
Orange Juice 10ml

and
Lemon Juice 25ml
Gomme (Sugar Syrup) 15ml
Blackberry Garnish

Add all ingredients to a cocktail shaker, with ice. Shake and then strain into a rocks glass - over ice. Garnish with a fresh blackberry.

Figure 4.17 The GPMO cocktail.

Actually, I personally think it is pretty tasty and goes down oh too easily!

Go with what feels right to you and I am sure you can come up with some amazing ideas for yourself. And don't forget to get your team involved and seek their creative thoughts – they will, I have no doubt, come up with some great suggestions (you just have to manage the budget 😊).

Have fun and build that team.

"Humour is by far the most significant activity of the human brain", Edward de Bono.[6]

Notes

1   The Project Manager Who Smiled: The Value of Fun in Project Management – Taylor, P – Publisher The Lazy Project Manager Ltd 2013.
2   You can check the book out to see what other PMOs have done for fun and team engagement.
3   Mentimeter is a Swedish company based in Stockholm that develops and maintains an eponymous app used to create presentations with real-time feedback. https://www.mentimeter.com/.
4   It was supposed to be the whole team, but a combination of travel visa issues, COVID, and general travel challenges, it was not to be. It was instead, most of the team. But we do aim to have our first all-team meeting in 2024 (fingers crossed).
5   My personal favourite was on top of the Marina Bay Sands hotel in Singapore – very special indeed.
6   Edward Bono was a Maltese physician, psychologist, author, inventor, and broadcaster. He originated the term *lateral thinking*, wrote many books on thinking including "Six Thinking Hats", and was a proponent of the teaching of thinking as a subject in schools.

Team building (communication)

You will build out your own cadence of team meetings and communication, but just for some reference this is what we do:

- Once a month, there is a full team meeting (virtual). Due to the global nature of the team and the costs involved, we have not had a full team meeting in a face-to-face format, but that said, everyone knows everyone else and communication can, and is encouraged, to take place at all levels. (We had one meeting with most of the team in the early days, but that was before we built out the Outcomes team fully, and even then, a couple of team members were unable to travel to join the meeting, as we noted earlier.)
- Weekly, I have a call with my three team leaders with the first call in the month being a formal initiatives update using a standard reporting template. All other calls in the month include whatever is "hot" for any of us to raise and discuss.
- The team leaders also have weekly calls with their teams and additionally they meet to discuss matters of alignment and priority. This is to ensure the maximum amount of integration and alignment.
- Each team has one physical team meeting each year to reinforce team spirit, plan, and review, and had some downtime to socialise and get to know each other even more.
- At my leadership level, I have a quarterly face-to-face meeting with my three leaders and sponsor, with the Q3 focused on our short-range planning/budget input and the Q4 being our year ahead plan-out in detail.
- Twice a year, I schedule a 1:1 with each team member – this is known as "skip a level" so that I have some direct engagement with every team member.
- And then, of course, there are the ad hoc meetings set up with any mix of the team on an as-needed basis.
- We also have a Teams chat channel set up.[1]

DOI: 10.4324/9781003346470-26

This works for us, although we have adapted this model a few times to improve team communication and alignment.

Now, do you know what makes a team really awesome?[2]

It's been a hot topic in the world of business and psychology for ages. There's been tons of research on team performance, from academics to big companies like Fortune 500. And get this, even virtual teams have been studied since the 1960s, so working remotely isn't as new as we might think. Some companies have been doing it for over 20 years. In fact, I have in my role as leading PMO on a global scale, always been working in this way.

Anyway, every team is different and goes through changes. Sometimes things get messed up and projects fall apart because of team issues, but there are some basic things that great teams have in common. You've probably heard of most of them, but let me go through a few key ones for high-performance teams:

- Everyone knows their roles, goals, and tasks clearly.
- There's a shared sense of purpose.
- People's well-being and work-life balance are taken seriously.
- Everyone feels like they belong.
- People take responsibility and hold themselves accountable.
- People take ownership of their work.
- There's room for autonomy.
- Learning never stops.
- Everyone is on the same page.
- The team can adapt quickly and be agile.
- Healthy disagreements are encouraged.
- Leaders are empathetic.
- People feel safe to express themselves.
- Mutual respect and trust are crucial.

Now, not every high-performance team will have all of these traits, as it actually depends on the team and the project/task they're working on. But it's important for all team leaders to know these traits in order to build out a high-performing team, which is what we all want of course, especially in the challenging world of PMOs.

Finally, as my team are all over the world, much of the communication is virtual in nature and also extends the working day for many to align the range of time zones we live and work in.

Our monthly team call is a great example – the "best" time we can select means that it is run 9pm to 10pm UK, 10pm to 11pm Germany, 4pm to 5pm US (East Coast), and 6am to 7am Australia.

For our stakeholders, we run sessions is that are regionally convenient for the "locals", meaning that the PMO team members often take the "hit" of running these sessions in times that are way outside their own normal working hours. It is all part of the job and being "global".

This was explored in "The Social Project Manager"[3] book and the three rules were:

• Be considerate of language.
• Be considerate of culture.
• Be considerate of time.

Whilst we operate as English being the standard business language, it has to be remembered that many people that we engage with are not native English speakers and this must be taken into account.

And, of course, we are working with many different cultures, so assuming everyone thinks and behaves the same as you is wholly wrong and will lead to many misunderstandings. One of the developments from the PMO has been a briefing guide for us, and our project managers,[4] on some of the key cultural differences that we should be aware of.

Then time, which we have covered above.

Don't always make the people you are working with engage with you in times that are not socially convenient to them – take the pain sometimes.

Notes

1   My team is doing their best to get me truly active within Teams (and JIRA), but I freely admit that I am a "dinosaur" when it comes to this stuff – although, I would argue, a highly effective and efficient one.
2   A summation of the thoughts in "Team Analytics: The Future of High-Performance Teams and Project Success" Goldman, B and Taylor, P – Publisher Routledge 2023.
3   The Social Project Manager: Balancing Collaboration with Centralised Control in a Project Driven World – Taylor, P – Publisher Routledge 2015.
4   Many of our projects and programs cover multiple countries.

We are all on a mission

And there you have it, the heart of our global PMO.

Project Community, Project Framework, Project Academy, and Project Life: These are the four pillars of our project world and the focus of all of our efforts to support our worldwide community of project managers and project leaders.

And, naturally, we have a mission statement:

The GPMO provides a centralised source of truth, enabling disciplined processes, and driving innovation, to ensure excellence in project delivery in supporting our global project community as deliverers of client vision.

Let's just break that down for a moment and replay it against the previously detailed focus areas of our PMO.

Centralised source of truth

What we mean by this is the PMO will be the central reference point or access for the current best practice project delivery through the project methodology of framework – which includes the optimum path to managing a project as well as the most up to date references, including artefacts and 'how do I' guides, – supported by a knowledge base of a global PMO and world-wide project community of professionals, – and further supported through a clear expectation of capability overseen by the project academy.

Enabling disciplined processes

Through the application of the project framework, supported and measured through the project health check program, a flexible but structured means to project success is laid out for all project managers to reference and apply – but as a living framework, open to continual improvements through periodic updates, these disciplined processes are regularly validated for optimum relevance.

DOI: 10.4324/9781003346470-27

Driving innovation

The PMO remains open to any technological or non-technological improvements in any and all of the services offered – the first two years of the PMO were predominantly foundational focused but the next year will be very much focused around automation and both accelerating project delivery whilst offering our project managers enhanced support through technology.

We also have an open policy to ideas generated from the field and have run an ideation program periodically to gather great ideas from those who deliver projects everyday – they are the true experts.

Excellence in project delivery

All the four components of our PMO world; Project Community, Project Framework, Project Academy and Project Life, are constructed for one primary reason and that is to deliver our customers the business benefit that our technology offers, and to do so in the most efficient and least disruptive manner that we can.

Deliverers of vision

Our ultimate goal is for our project managers to not be technical task masters, ticking boxes and filling in status reports and the like but to be elevated to the status of deliverers of the vision that our customers desire.

Part 5

Success and celebration

In Part 5, we consider what success looks like, how we measure progress against success, and how we (regularly) celebrate success, both within the team and within the global project management community.

We conclude with the important mantra of 'always be learning'.

DOI: 10.4324/9781003346470-28

Measuring progress

> BEST PMO TIP: The best PMOs always measure progress and adjust as needed based on feedback.

In "Delivering Successful PMOs", I outline the steps and strategies needed to design and deliver the best project management office (PMO) for your business, now updated in this book with the "Outcomes" focus.

One key takeaway is the importance of measuring the PMO's value regularly.

Many organisations do not consistently evaluate or measure the success or returns on investment (ROI) of their PMO. By measuring the PMO's value, a PMO leader can articulate its true value to the business and continuously improve its performance.

In lesson 4, the book emphasises the need to "lock in" the value of the PMO during the delivery period and regularly reassess and measure it. The value of the PMO should not be taken for granted, and a good PMO leader should continually demonstrate its value to stakeholders. The guide argues that there is a direct link between the maturity of the PMO and the value it provides. Mature PMOs are far more likely to offer a real competitive advantage to a business by increasing the speed and quality of business returns.

The book also encourages PMO leaders to learn from their experiences and avoid repeating the same mistakes, suggesting that in project management, doing the same thing over and over again and expecting different results is insanity. Well, to be honest, it is true in all aspects of life, unless you believe in magic.

By continually assessing and measuring the PMO's value, a PMO leader can identify areas for improvement and adjust their approach accordingly.

DOI: 10.4324/9781003346470-29

All the above is very much linked to the thoughts shared in the chapter on "Adoption and Change" in Part 7 of this book, so at this point I will say no more, bar asking you, right now, to consider how you are measuring progress against the original plan and current need and if you are, how well are you doing that measuring, how objective is the assessment, how inclusive of all your stakeholders is it (honestly)?

And if you aren't measuring, then start today!

In my own PMO's case, we reported out 90 days after I joined in a "State of the Nation" summary, and then again 180 days after joining, and then once more one year after joining – all in some detail showcasing what we had achieved and what we had left to do.

I look forward to preparing the second anniversary report in a few months from now. I know it is going to be a good one.

The Global PMO now reports to a body of SVPs known as the "Services Council" and, once again, the PMO focus and associated successes are regularly assessed against the business needs defined by this collective group, based on a combination of strategic objectives and more tactical needs.

Celebrating success

BEST PMO TIP: The best PMOs are never quiet about achievements and successes but shout out loud to thank contributors and showcase progress.

Celebrating success, achievements, progress, milestones, etc., is an important part of building an effective, self-assured team, and making your PMO team (or any team) a great place to be a part of.

Why is it a good thing? Well, it can significantly contribute to:

- Facilitating team building – especially if you have a newly formed team
- Fostering a positive environment – reinforcing the achievements and successes
- Enhancing team/company culture – the team should reflect the company culture, but you can take it to the next level
- Instilling a sense of appreciation among employees – the cycle of achievement, reward, and more achievement
- Boosting productivity levels – driven by the sense of appreciation and belonging to the team
- Strengthening communication channels – recognizing people and teams, sharing the recognition, and having fun
- Increasing employee retention and job satisfaction – building brand and team loyalty
- Cultivating a positive PMO team reputation – attracting support and, potentially, new team members
- Stimulating creativity and innovation – creating a safe place for ideas and creation

There are many ways to celebrate achievement. The key is to understand your team members and what motivates them.

DOI: 10.4324/9781003346470-30

We have celebrated in a number of ways:

- A simple "thanks" on team calls
- Using the formal company recognition and reward program
- Spot bonuses
- Promotions and salary increments
- Share options
- Small "gift" awards
- Posting "thanks" on communication channels
- Delivering "shout-outs" during our various events

It doesn't have to be big, but it has to be sincere, and it does have to happen.

"Success is best when it's shared." Howard Schultz[1]

Note

1   Howard D. Schultz is an American businessman and author who served as the chairman and chief executive officer of Starbucks.

Always be learning

BEST PMO TIP: The best PMOs adapt to the enterprise's strategic expectations and know how to operate effectively within the corporate structure and culture. And they are not rigid in their own structure and focus in order to adapt and adopt quickly.

"Always Be Learning": A New Model for Lifelong Education, published by *Forbes*,[1] says it all.

Learning should be a lifelong endeavour, not something that stops at age 18 or age 22. People who continually upgrade their skills are not only better employees, but happier and more fulfilled people in general.

Now that is true for individuals, and it is equally true of teams – PMOs included.

The tip above says it all really. Whilst you, the PMO, are delivering change, there will be other changes happening all around you that will impact your work, like small (or sometimes large) gravitational forces, and then the goalposts will almost certainly move – maybe not radically but they will move. Whatever your original road map says, what is required at any point will be some variation of that plan. So you need to keep learning about what is relevant now and find that critical balance between your own PMO strategic vision and the needs of the business today.

So, the message is clear – keep on learning about your organisation's needs and priorities and adjust accordingly. This is covered a whole more in the "Adoption" chapter.

And associated with that as a member, or leader, of a PMO always be learning – from others, from books, from events, from your peers, from your stakeholders. In fact, any way you can learn, then learn. And give your

DOI: 10.4324/9781003346470-31

team the freedom to learn in the manner that works for them and in the areas that really interest them; it will provide a huge return on investment.

The old saying *"What if I train them and they leave?"* and *" What if you don't and they stay?"* by W. Edwards Deming[2] is a clever one of course[3] and if you substitute "train them" with "allow them" to continuously be learning brings it somewhat up to date. I go back to the argument for building the best PMO with the "right people". You secured (I hope) the best people to join your PMO so keep them at their best through freedom to learn, with your team or outside your team.

Personally, for me, attending one big project management conference a year and being challenged, inspired, and educated by some great speakers is a huge part of my personal learning. The rest comes from my team and the knowledge that they have, together with their noble attempts to teach this old dog (dinosaur) new tricks.

Notes

1   https://www.forbes.com/sites/schoolboard/2016/03/16/always-be-learning-a-new-model-for-lifelong-education/?sh=62a55f2a46bf.
2   William Edwards Deming was an American engineer, statistician, professor, author, lecturer, and management consultant.
3   We saw variations on this earlier with Ford and Branson's quotes.

Part 6

The launch

In Part 6, we outline the culmination of this PMO journey to date, the global launch event called, genius branding here,[1] 'The Event'.

There is still much left to be accomplished so consider this but a major milestone – after all, there is a two-year road map for delivering even greater value to the business so much is left to be done, and the PMO work goes on.

The following describes the planning, the buildup, and the delivery of this key event, along with lessons learned through this experience.

Note

1 I will be honest here, not all of my team immediately reached the conclusion that calling our event 'The Event' was marketing brilliance, but I believed in it and, personally, I think it worked.

DOI: 10.4324/9781003346470-32

The cunning plan

The purpose of what became known as "The Event" was:

- To accelerate and enhance project manager enablement and effectiveness
- Continued improvement in managing customer and project activities to achieve scoped hours/cost/margin
- Advance the supporting methods and tools to enhance PM capability and standardised project delivery
- Advance a more consistent, scalable project quality/health check process

Now how do we do this most effectively for a group of 150 plus project managers, all of their managers (and their managers), as well as a large community of consultants, in many countries around the world?

How do we have a real impact and make this stand out from all of the other meetings and communications that this same community received?

And how do we get maximum buy-in (adoption) of all the changes that the PMO was promoting?

Well, make it big and make it exciting!

The PMO had a cunning plan.

Eventbrite[1] noted *"Events are one of the most powerful marketing tools because they center around an intentionally designed experience. And experience is in high demand. According to Eventbrite's research, 77% of millennials say some of their best memories are from an event or live experience they attended"*.

Our own experience when we have run previous events such as our PM Summits (virtual) is that most people love it, and therefore the "event" itself was designed to be a bigger version of this and gain interest and "hook" our global audience of project managers.

We succeeded.

DOI: 10.4324/9781003346470-33

Note

1 Eventbrite is an American event management and ticketing website. The service allows users to browse, create, and promote local events. The service charges a fee to event organisers in exchange for online ticketing services unless The Event is free. https://www.eventbrite.com/

The creation of "The Event"

OK so a little bit of honesty here.

We had a cunning plan A and then we had to switch it to a cunning plan B, and then adapt that plan to cunning plan B2.

The original ask of the business was three regional face-to-face summits over a minimum of three days to launch a number of initiatives and to train attendees in the relevant material (and have a really good party of course).

As the budget ask was just too steep for the business (which we always suspected might be the case), we moved (as nimble and agile as you could ask) on to a virtual event supported by just-in-time training and ongoing support.

The actual plan was a three-hour virtual delivery showcasing the amazing work the PMO had led in the development of an all-new project and program flexible framework for our delivery project managers around the world. Not just that, but a launch of the ground-breaking Project Academy would allow our project leaders to have clarity of expectations in each role, clear guidance for career progression, and offer up public badging of a number of key skills that were required to be project managers for our business.

Plan B2? I hear you ask. Well, that was a small adaption in one region to move to hosted events (that is, PMO members attending and leading small local communities watching The Event recording and then discussing live).

For the rest of the world, it was live, centrally hosted, with presenting hosts, and a series of recorded sessions, live interviews, and live polls.[1]

As we wanted to showcase all of the GPMO team in some part it was decided that it would be best to record some of the content for a seamless transition and a high-quality delivery – it all helped with understanding The Event timeline. Plus, the team had a huge amount of fun

DOI: 10.4324/9781003346470-34

recording their sessions with a professional video team (some for the very first time).

Note

1   And some high-energy music – we elected to use "It's My Life" by Bon Jovi as our main theme tune.

The buildup

Having decided on "The Event", it was then critical to build interest over a five-month period of time. Give hints, entice, intrigue, and get The Event in people's calendars.

This was done in four ways:

- Offering up teasers in our monthly pulse sessions
- Sending our "save the date" invitations
- Getting senior management to "push" the "event" at every opportunity and ensure attendance
- Direct engagement with PMs and PM leaders to encourage attendance

Personally, I had a lot of fun designing some awesome promo videos using tools such as Flexclip https://www.flexclip.com/ – give it a go yourself. The movie trailer was a big hit, and I found the tool easy to customise and delivered shockingly good output.

I also used the stunning combination between ChatGPT (script creation) and Synthesia, an AI-powered video creation platform (check out the multi-language capabilities) https://www.synthesia.io/ – this creates pretty realistic avatar videos.

We also promoted that we had a special internationally recognised (bit of a superstar) external speaker who would talk about the future of project management (see Appendices for speaker recommendations).

The key here is to keep people interested. Entice, don't bore; intrigue, but don't give it all away before the occasion itself, and make people want to know more. Be creative and be different.

DOI: 10.4324/9781003346470-35

Chapter 6.4

The outcome

The Event was just the opening gambit; the beginning, the start. It was all about showcasing the new offerings and letting people know the PMO was there to support them in their future project careers and project delivery.

What follows is a just-in-time training approach for everyone and an ongoing "hyper-care" style support for project leaders in their day-to-day challenges as well as feedback on the new framework and academy so that we can make it even better based on practical application.

But one thing is for sure – we created a global buzz with our approach and our content, and right now we are planning 'The Event 2024'.

DOI: 10.4324/9781003346470-36

Chapter 6.5

Do you need an "event"?

Well, only you can know that.

But I would encourage you to consider something that is beyond the norm and attracts your project managers to attend and join in some form of collective celebration.

Keep it interesting for all your stakeholders and help break down operational or regional silos as well as put your PMO "on the map".

Don't be afraid to be different.

DOI: 10.4324/9781003346470-37

Part 7

Adoption and change

In Part 7, we address the critical theme of adoption is explored. Adoption of the changes that the PMO is aiming to bring about and, by default, the adoption of the changes (or transformations) that the organization demands.

We conclude with a look at progress tracking and reinforcement of objectives.

DOI: 10.4324/9781003346470-38

Change is hard

BEST PMO TIP: The best PMOs have a good understanding of the principles of change and the best approaches to change management (and adoption).

This is a topic I explored in my book "How to get fired at the C-level: Why mismanaging change is the greatest risk of all".[1]

It can be frustrating when change doesn't happen as expected in your organisation, despite multiple business cases being reviewed and approved, programs and projects being sanctioned, and funds being released. It's worth considering why change is challenging and often resisted.

Every project within your organisation, regardless of its size, is a change, and people generally don't like change. This applies to you, your colleagues, and everyone else. However, living in a world of constant change means that it's necessary to reach a point of conscious decision to make a change, especially when directing an organisation in a particular direction, achieving new growth, tackling new markets, releasing new products, expanding horizons, complying with regulations, and so on.

Although change in an organisation may be necessary, each change impacts individuals, making it a complicated process. It's human nature to resist change, but it's also essential for business objectives.

In considering change, you can describe it as a dynamic of where you currently are (C), where you want to be (D), and the resultant benefit for any change undertaken (B^2). For example, "My house feels crowded and noisy, and on top of that I have nowhere to put things … And this is making life less pleasant!" can be described as "My current house is too small (C), I would like a bigger house (D), the benefit of having a bigger house would be more room for myself and my family (B)".

DOI: 10.4324/9781003346470-39

When it comes to change, there are various states in which you may find yourself. You may lack insight into your problems or need to change to bring about some form of life change, have insight but need a solution or plan, or have a plan but need some assistance in making it happen. To move toward change, you need to balance various factors that may prevent you from doing so. Resistance to change can be caused by cost, risk, pain, or hidden reasons. Hidden reasons are often challenging to uncover and quantify, making it harder to assess the balance of resistance.

On the other side of the balancing scales are the reasons for change, including needs, problems, benefits, and implications. For change to happen, the scales need to fall more heavily on the side of "for change," ensuring that change is a reality. The Formula for Change,[3] created by Richard Beckhard and David Gleicher, and refined by Kathie Dannemiller, helps model the process of change. The formula $(D \times V \times F \times CL > R)$ states that dissatisfaction with the status quo, a vision of the future, first steps toward change, and the cost of change must be greater than resistance to change for change to occur. Therefore, it's crucial to consider the dynamics of change, the states of change, and the balancing act necessary to make change happen.

The key here is that for anything that your PMO is aiming to deliver, then the advantages are often easy to calculate but the value to those impacted are less easy to appreciate sometimes.

- Needs - The definable drivers for adopting a process of change, the need of the person or persons to make a change
- Problems - What is it that is causing some issue or concern in the status quo that offers the desire to make some form of change?
- Benefits - What are the desirable benefits of such change and the expected beneficial outcomes of adopting something new?
- Implications - If no change is initiated, then what will the impact be and what will the consequences be? The implication is that something must be encouraging the need for change in the first place.

Therefore, you need to make sure that the scales fall more heavily on the side of "for change" in order to stand a chance of making such change a reality.

Notes

1   How to get fired at the C-level: Why mismanaging change is the greatest risk of all - Taylor, P - Publisher TLPM Publishing 2017.

2   C stands for Current, D stands for Desire, B stands for Benefit.
3   The original formula, as created by Gleicher and authored by Beckhard and Harris, is: $C = (ABD) > X$ where C is change, A is the status quo dissatisfaction, B is a desired clear state, D is practical steps to the desired state, and X is the cost of the change. It was Kathleen Dannemiller who dusted off the formula and simplified it, making it more accessible for consultants and managers.

Adoption is harder

At its most fundamental level, change involves moving from the current state to the transition state and ultimately, the future state. It is a necessary process for any organisation that wishes to stay competitive and relevant in an ever-changing market. However, simply managing change through these phases is not enough; it is crucial to adopt change and make it an integral part of the organisation.

Change adoption is critical to ensuring that new initiatives have the maximum impact on supporting strategic objectives, motivating the organisation, and achieving differentiated results. Without change adoption, new initiatives may be met with resistance or indifference, leading to limited success, and wasted resources.

Rolling out a new program or project is one thing, but ensuring its adoption is another. For a change effort to be successful, the leadership team must answer several questions. How can they make the change stick? How can they measure and demonstrate tangible results? How can they continue learning and evolving?

Effective communication is one of the key tasks involved in change adoption. It is crucial to communicate the change clearly to all stakeholders, including employees, customers, and suppliers. Effective communication involves not only informing stakeholders of the change but also helping them understand why the change is necessary and what benefits it will bring.

Creating a sense of urgency is another critical task involved in change adoption. Urgency is essential because it encourages people to embrace the change and act. It involves demonstrating why the change is necessary and what will happen if it is not implemented. Without a sense of urgency, people may be less likely to take the necessary actions to adopt the change.

Developing a well-planned implementation plan is also critical to successful change adoption. Change should be well-planned to minimise disruptions to business operations. A plan should be developed that

DOI: 10.4324/9781003346470-40

outlines the steps required for successful implementation. It should also identify potential roadblocks and strategies for overcoming them.

Identifying and addressing resistance is another important task involved in change adoption. Resistance to change is common, and it's important to identify and address it early on. This involves understanding the reasons for resistance and developing strategies to overcome them. By addressing resistance, organisations can increase the chances of successful change adoption.

Providing training and support is also critical to ensuring successful change adoption. Employees may need training and support to adapt to the change. This could involve providing training sessions, coaching, and other forms of support. By providing training and support, employees can feel more confident in their ability to adapt to the change.

Monitoring progress is also essential once the change is implemented. Regular monitoring helps to identify any issues or challenges and adjust as necessary. It is also important to celebrate success. Celebrating success is an important part of change adoption because it helps to reinforce the benefits of the change and encourages people to continue to embrace it.

Resistance to change can take various forms and understanding the reasons for resistance can help organisations address them effectively. Some common types of resistance include fear of the unknown, comfort with the status quo, lack of trust or credibility, personal impact, inadequate communication, and lack of skills or knowledge. By addressing these types of resistance, organisations can increase the chances of successful change adoption.

A fuller list is here:

- Fear of the unknown: People may resist change because they are uncertain about what the change will bring and fear the potential negative consequences.
- Comfort with the status quo: People may prefer the current way of doing things and resist change because it disrupts their comfort and routine.
- Lack of trust or credibility: If people do not trust the source of the change or the process used to implement it, they may resist it.
- Personal impact: People may resist change if they perceive that it will negatively impact their job security, status, or work-life balance.
- Inadequate communication: If people do not have a clear understanding of why the change is necessary, or how it will be implemented, they may resist it.
- Lack of skills or knowledge: People may resist change if they feel they do not have the skills or knowledge required to adapt to the new situation.

- Understanding the reasons for resistance to change can help organisations address them effectively and increase the chances of successful change adoption.

In conclusion, change adoption is critical to the success of any change effort. It involves several key tasks, including effective communication, creating a sense of urgency, developing a well-planned implementation plan, identifying and addressing resistance, providing training and support, monitoring progress, and celebrating success. By adopting these tasks and understanding the reasons for resistance, organisations can increase the chances of successful change adoption and achieve their strategic objectives.

Chapter 7.3

Adoption
The overlooked side of change

[Peter: The following section on Adoption was written by David Ayling-Smith]

In today's rapidly evolving world, where organisations face constant disruptions and challenges, the ability to navigate and embrace change has become critical for business survival and success. Change adoption, the process of driving individuals and organisations to willingly embrace and internalise transformation, plays a crucial role in optimising the impact of new initiatives, achieving strategic objectives, motivating the organisation, and driving meaningful change. This chapter explores the concept of change adoption, highlighting its importance, challenges, and strategies for success.

Change adoption goes beyond the traditional concept of change management, which primarily focuses on managing the process and mechanics of change. While change management ensures a structured approach to implementing change, change adoption places emphasis on driving individuals and the organisation as a whole to genuinely embrace and integrate the change into their mindset, behaviours, and culture. It recognises that change cannot be forced upon individuals; rather, it requires their willingness, commitment, and active participation.

All projects by definition are change management activities. Whether an organisation is going through a wholesale transformation leveraging new technology to reorganise their business processes, or a smaller scale software replacement, success or failure will depend on the attention given both to the technology and people elements of the project. The human side of adoption brings into play a panoply of human emotions, from fear, doubt, and derision to enthusiasm, optimism, and hope. Managing and motivating people to change is very different from the technical planning process and requires preparation, engagement, and education with a particular focus on adoption to harness all those positive human elements and act to minimise potentially negative behaviours.

DOI: 10.4324/9781003346470-41

Resistance to change is one of the most significant challenges faced by organisations during the adoption process. Several factors contribute to resistance, including comfort with the status quo. Many individuals are content with their current practices and perceive them as efficient and productive. Learning a new way of doing things and unlearning old habits can be disruptive and unsettling. Additionally, the fear of the unknown and uncertainty about the outcomes of change often create resistance among individuals.

Another critical factor influencing change adoption is trust. When change is imposed upon people without their input or understanding, it can breed scepticism and cynicism. Employees may view the change as just another corporate initiative without perceiving its relevance to their daily work lives. Building trust requires transparent communication, actively involving employees in the change process, and clearly articulating the purpose and benefits of the change. Addressing individual concerns and demonstrating a genuine commitment to employee well-being can help overcome resistance and foster skills and capability gaps that can also hinder change adoption. Individuals may resist change because they believe they lack the necessary skills, knowledge, or capability to adapt successfully.

Organisations must provide adequate support, resources, and training to facilitate the learning and development required for change adoption. Offering coaching, mentoring, and continuous learning opportunities can empower employees and boost their confidence in embracing the change.

Without people engagement, there can be no change, and yet people change for different reasons and at different speeds and this needs to be considered during a successful project. There is much in the literature that deals with these themes, and often from diverse fields of study.

Maslow's Hierarchy[1] of Needs explores what motivates us and, when translated into the field of adoption, can be represented by the highly engaged response of "I'm into this", through to the highly disengaged response of "I'm going to lose my job".

Kübler-Ross's[2] 5 stages of grief can be used to explore adoption pathways with its description of denial, anger, bargaining, depression, and then integration into the new normal.

The theoretical learning model[3] describes how we move from unconscious competency and dip into conscious incompetence as a transformation takes root and then the slow climb through conscious competence and then back to unconscious competence.

The ADKAR[4] acronym also helps us understand the dynamics of human changes with its guidance on:

- Awareness – What is changing
- Despite – Why is this good for me
- Knowledge – Help me understand

- Ability – Support me to operate in the new way
- Reinforcement – Embed the change

Individuals need to understand how the change is connected to their roles, responsibilities, and the overall purpose of the organisation. Communicating the impact and benefits of the change in a relatable and meaningful manner can help individuals recognise their personal stake in the transformation. This relevance can manifest as individual benefits, such as increased efficiency, more engaging work, and career growth opportunities, or as organisational benefits, such as improved performance, competitiveness, and sustainability.

All this messaging of course requires a laser focus on communication and there are many tips and tricks that have been used by practitioners over the years that help improve the efficacy of the message.

Focusing on the why of the change as much as the what. A common mistake made during people-centric change is a mechanistic approach to training and objectives, whereas inevitably people want to understand why, a higher-level positioning strategy is often all that is required to ignite people's understanding and interest. Talk as much as practical about what is "in it" for the individual, understanding is important, but motivation is critical if change is going to be adopted. A helpful approach to communication change that leads to better adoption is to take people on a journey, tell the story of the transformation in the language of the listener, keeping simple and relevant and accessible to maximise the chance of being heard. To further improve the likelihood of the message being reviewed, high-frequency, low-amplitude communication is helpful. Said another way, communicate little and often in a predictable way that the team comes to rely on as a source of truth. Regular communication like this will build trust, provided it is honest and that means also sharing things that have not gone well. Without this transparency and openness, adoption is vulnerable to being eroded by cynicism and disengagement.

This focus on people adoption needs managing as rigorously as any technical project. Preparation is key and starting with the end benefits in mind is a helpful technique. Having an agreed and communicated value statement helps subsequent measures of benefit realisation. But there are more prosaic early measures of success. Woody Allen the famous film director claimed that "80% of success is just showing up" so measuring usage of the new system/ process/tool can provide early-warning indicators. Pulse surveys can also be a handy technique to assess engagement but ultimately there need to be adoption targets that get measured and constant interrogation of barriers and understanding of blockers and issues that need to be addressed to maintain momentum. In order to be able to demonstrate progress, early wins are useful proof statements of success.

It has been demonstrated that success fuels momentum by supporting morale and generating energy and enthusiasm for what is often a long slog. Of course, to have early wins it takes us back to the importance of preparation and savvy planning to ensure that early goals are achievable.

However, all people-centric adoption will be characterised by a productivity gap at some stage of the project. The period where the team is using the new system/process/tool but are not yet experts means they will make mistakes. This is one of the causes of the despair stages encountered on a transformation journey and it reinforces the need for motivation provided by early wins and to break down the work into manageable chunks. It is almost always ill advised to try and climb the mountain in one go. The context for the change is also relevant here, as we look at maturity models and recognise that in a world of continuous transformation we are always somewhere on the curve from good, to better, to best. It is important to recognise where you have come from "This isn't perfect, but it is way better than what we had before", through to re-stating what "better" means as you transform. Technology has always been a powerful lever for change, and it is often not apparent what is possible until you are in the middle of the transformation.

Alongside all the positive management techniques to help drive adoption, be vigilant and aware of the many inhibitors for adoption and respond quickly and purposefully as you recognise them in your project. Here are a few.

Engaging stakeholders at all levels in the organisation and involving them in the change process is vital; it fosters a sense of ownership and commitment. Actively seeking their input, ideas, and concerns enables a more inclusive approach and encourages their active participation in the adoption journey.

Additionally, identifying and empowering change agents within the organisation can significantly impact change adoption. Change agents serve as advocates, influencers, and role models who can inspire and support others in embracing the change. They should be equipped with the necessary knowledge, skills, and resources to effectively drive the adoption process.

Sponsorship is essential for a fast-moving transformation and lack of exec sponsorship is one of the top three reasons projects don't fulfil their potential. Policy changes or technology enabled possibilities need to be evaluated and agreed quickly so that the project doesn't stall, or the adoption phase lose momentum. Additionally having a leadership engagement in the communication of the whys and wherefores of the endeavour help re-enforce the importance and relevance to everyone taking part. Of course, the reverse is true, and many organisations find themselves engaged in multiple transformation activities at the same time and this can rapidly lead to fatigue and distraction as subject matter experts and key employees find themselves unable to deal with the complexity of constantly moving objectives. The answer to this phenomenon is to again plan carefully and be thoughtful about the organisation's ability to absorb the changes proposed.

The persistence of an old system/process/tools is also an indication of poor adoption. There are lots of theories on how to deal with this from the "burn the bridges" approach and physically removing access to old systems or technology or the gradual cutover technique. Both have their advantages and issues but, in all cases, appropriate high-quality training and sold planning can ameliorate the issues.

Finally, adoption is a long game and often this work doesn't start until after a technology project is live, when people are fatigued and relaxing after the euphoria of a successful launch. But this is precisely when many of the techniques already discussed need to be put into place. Cultivating a culture of continuous learning and growth is essential for change adoption. Encouraging employees to embrace a growth mindset and providing opportunities for skill development and knowledge sharing create an environment where change is seen as an opportunity rather than a threat. Celebrating and recognising individual and team successes along the change journey reinforces positive behaviours and encourages continued commitment.

Change adoption is a critical component of successful organisational transformation. By understanding and addressing challenges such as resistance to change, lack of trust, skills and capability gaps, and perceived relevance, organisations can optimise the impact of their initiatives and create a culture that embraces and thrives on change. Through clear vision, effective communication, stakeholder engagement, empowerment of change agents, provision of support and resources, fostering a culture of continuous learning, and monitoring progress, organisations can navigate change successfully and achieve their strategic objectives. Ultimately, change adoption empowers individuals and organisations to adapt, innovate, and thrive in an ever-changing world.

Notes

1   Maslow's Hierarchy of Needs is a theory of motivation which states that five categories of human needs dictate an individual's behaviour. Those needs are physiological needs, safety needs, love and belonging needs, esteem needs, and self-actualisation needs.
2   Elisabeth Kübler-Ross was a Swiss-American psychiatrist, a pioneer in near-death studies, and author of the internationally best-selling book, "On Death and Dying", where she first discussed her theory of the five stages of grief, also known as the "Kübler-Ross model".

3   Theoretical learning models provide a framework for understanding how individuals acquire knowledge, skills, and understanding.
4   The ADKAR Model of Change Management is an outcome-oriented change management method that aims to limit resistance to organisational change. Created by Jeffrey Hiatt, the founder of Prosci, the ADKAR Model is the Prosci change management methodology.

Chapter 7.4

Building the portfolio dashboard

BEST PMO TIP: The best PMOs have clear visibility into the progress and cost of all projects. They also know exactly how resources are being used. They openly share this information with all the appropriate stakeholders throughout the enterprise.

I assume everyone understands portfolios and dashboards in principle but, naturally, everybody probably understands something different.

We are using our portfolio dashboard for three purposes:

- Marketing – What does the PMO do?
- Controlling – When is who doing what?
- Collaborating – Working together to find a solution.

We have been on a journey with this, similar to the team evolution. We had to come up with different approaches and looking at the current one, we are right at the start.

You may ask, why? It's as simple as this: no approach fits all.

Looking back in history, the earlier PMO team was just three people based in the United States, easy to collaborate and exchange information, with everybody being aware of who was doing what – in fact, everyone did pretty much the same thing, jumping in to help out when needed.

Moving on, as the team became bigger, but still mainly based in the United States, they had to start using a type of centralised repository to keep their work and results updated. But now, with a team size of around ten people, they still could work well as it was simple to get on a call to discuss things.

It got interesting as there was a need to establish a strict intake process at the same time the team became global. There was no chance for everyone to

DOI: 10.4324/9781003346470-42

jump on a call to discuss something urgent anymore. That's when emails came into play, and when we say emails, it was a lot of emails. So, the decision was made to adopt a tool that at a minimum kept track of who was doing what with a level of breakdown of major initiatives.

Well, early days, but no one really updated the status or progress, and as we reached the current team size, especially with the build of the three teams (P:M:O) being truly global – we simply had to evolve. But that did not happen overnight. We had to go through some pain points with a very steep learning curve first.

Now we are using one tool for controlling and collaboration. Everybody can see who is working on what and even make comments on tasks. But honestly, emails are still flying around.[1]

Note

1 Peter – OK I admit it I am old school, a dinosaur, but am awesomely responsive on email but … .

Driving the KPIs

KPIs[1] are important to either monitor progress or to identify issues and challenges. But to make it very clear, not all KPIs work for every organisation. Therefore, you should not incorporate KPIs you have seen from other organisations just to have KPIs in place. It won't work. Also, too many of these might drive you into a wrong direction and let you miss the root cause of any issues or deviations.

So it's on you to come up with the right KPIs for your organisation – typically tough work, but critically important for future success. And don't forget, you need to define any limits or bandwidth for each of them as well.

Here are some examples we have developed and are currently monitoring to see if they are working for us or more importantly if they provide us with an indication of an upcoming issue ahead of time to be able to act vs. react.

- Proportion of possible net project management hours to assigned and unassigned PM hours over the next one, three, and six months.
- Proportion of active projects and current phases to number of project managers and scheduled Kick-Offs and Go-Lives.
- Total value of the company based on our project management scorecard with trends over months.

All numbers can be aggregated to the organisational top level, but more importantly, you can drill down to each region and even to each delivery group. And this is where your limits and bandwidth need to be located. If the aggregation shows green, it does not have to be on every level below.

KPIs offer an understanding of your business, but they need to be validated every year and maybe adjusted as the business progresses.

DOI: 10.4324/9781003346470-43

Note

1   KPI stands for key performance indicator, a quantifiable measure of performance over time for a specific objective. KPIs provide targets for teams to
shoot for, milestones to gauge progress, and insights that help people across
the organisation make better decisions.

Chapter 7.6

Winning the hearts and minds

At the end of the day, it is all about providing value and winning both the hearts and the minds of all those impacted by the work your PMO undertakes.

We've all heard the saying that if we want to persuade others successfully, we need to win both their hearts and minds. I explored this in my own book, "Project Branding", some time ago.

The tricky part is that winning hearts and minds often feels something of a paradox and since people are generally complex, as are the problems you will be trying to solve no doubt, attempting to appeal to both emotions and logic can actually make us less influential if we don't have a solid plan.

Winning Hearts

When it comes to persuading by winning hearts, it's all about emotionally connecting people to your idea or position. In any persuasive conversation, you need to establish some level of connection with others.

The best way to persuade in these circumstances is to connect with people on a personal level. Think of it as creating a "hook" that grabs their attention. Use vivid descriptions and metaphors to draw others into your vision and share personal stories and experiences to demonstrate that what you're suggesting is (you strongly believe) the right choice or decision. It is about making a good connection with people.

Winning Minds

When it comes to the science of persuasion, winning minds means using logical and well-articulated arguments in favour of your proposition. If you're trying to persuade anyone about anything, you definitely need a logical argument to support your perspective.

To win over people's minds, it's important to carefully craft your message (back to marketing again). Find common ground with your

DOI: 10.4324/9781003346470-44

audience and share your expertise and understanding of the issue, high-lighting any analysis you've conducted, and providing evidence to support your position. Present the benefits in practical and tangible ways.

Hearts and Minds

The paradox of persuasion doesn't have to hinder our ability to influence others effectively. It is crucial to pay close attention to your audience and the task at hand. Identify the strongest approach (the balance between hearts and minds) based on the situation as you understand it.

You can't win the hearts and minds of the masses unless you inspire them – you must lift their spirits and enliven their hearts Jason Silva[1]

Note

1   Jason Luis Silva Mishkin is a Venezuelan-American television personality, short filmmaker, futurist, and public speaker. He is known for hosting the National Geographic documentaries *Brain Games and Origins*.

Part 8

Reflections

In Part 8, the author reflects on the journey and the most important aspects that have led to the 'now' and the level of success achieved by the global PMO team.

DOI: 10.4324/9781003346470-45

Time flies

As noted in the introduction to this book this reflects a practical journey of 24-months designing and then building a Global PMO for today and for tomorrow, focusing on "Projects", "Methods", and "Outcomes".

> BEST PMO TIP: The best PMOs (I strongly believe) take time to reflect on the past as well as the future.

The structure is a real-life case study and, what I hope you will see, practical models to follow or guide you to build or improve your own PMO. It was definitely designed as a "dip in and out" sort of book, but also follows a logical flow of PMO construction and purpose.

The path to this point has seemingly flown by. It truly seems like it was only the other day I began conversations to start to build a road map for this particular PMO.

But, also noted, whilst many foundational initiatives have now come to fruition the journey continues.

We have developed a two-year road map for what comes next and what then comes after that, providing a strategic vision supporting the organisation's goals, whilst also being ready for any tactical urgent needs from the business.

All whilst being cognisant that life changes and our organisation may well need us to take a different path, in which case we will pivot and move to those revised objectives and purpose.

DOI: 10.4324/9781003346470-46

Chapter 8.2

Key contributions

I would, in particular, speak to three key, even critical, contributions to the success that my Global PMO has enjoyed, and, I believe, my company has, in turn, benefited from.

The first is a huge shout-out to the executive leaders who have strongly believed in and sponsored the PMO – without this, as we noted in the "Adoption" chapter, there wouldn't be a PMO at all, and certainly not one of such sophistication and scale that I have had the privilege of leading.

Secondly, the "Projects: Methods: Outcomes" construct completely changed this approach of the PMO and provided a real business-focused team to oversee the delivery of value to their organisation. Through this formation, we have really bridged the gap being the PMO and the project managers on the ground, delivery every day, and overcoming challenges every day. They have the knowledge and the wisdom; we need to be the custodians of such and distribute it for the common good.

And thirdly, my amazing team – see the Wall of Fame in the appendices.

As I noted in my book "Leading Successful PMOs", and something I believe even more today. *"The PMO is all about doing it all but with the right team*[1] ...".

I certainly have the "right team" and could not have achieved what we have achieved without each and every one of them.

Note

1 See Appendices for my own "keep it simple" definition of the world of a PMO.

DOI: 10.4324/9781003346470-47

Chapter 8.3

Sponsorship is all

As described, right at the start of this book, the following are the top challenges for any PMO:

1   Lack of executive support: PMOs require executive support to be effective. Without support from senior leadership, it can be difficult for PMOs to get the necessary resources, authority, and buy-in from stakeholders.
2   Inadequate funding and resources: PMOs require adequate funding and resources to deliver value to the organisation. Without sufficient resources, PMOs may struggle to establish the necessary standards, processes, and governance.

I was lucky. My team was lucky. We had such sponsorship[1] in full, and we have had all "reasonable" requests for resources supplied (that is all those supported by a strong and clear business case).

So much more for me to say at this point, except please do make sure you have solid sponsorship for your PMO and that is stays "solid", by closing the loop on what is going on and what is required (from your sponsor(s)' perspective).

In my case, this individual sponsorship was eventually replaced by a formal governing body we refer to as the "Services Council". All initiatives, PMO, and others, go through this group for review, validation, and sanction. It works.

Note

1   I am taking it as a strong indication that nothing has changed as my original sponsor wrote the foreword to this book.

DOI: 10.4324/9781003346470-48

It all takes (more) time

Now, generally speaking, I am an impatient "let's get this done" sort of person and the PMO has pushed some things out in a "fast and furious" style, but most we haven't.

Taking people on any journey with you takes time in discussing, sharing, communicating, receiving feedback, and balancing priorities so do not be surprised that your initial timeline is rapidly lost and replaced by a more realistic one (see the "Adoption" chapter for me on this).

One of my mentors noted that I *"was the master of under promising and overachieving"* and there is nothing wrong with that, I say, but make sure that it all fits under a clear strategic vision.

DOI: 10.4324/9781003346470-49

Making a difference

In another book, my co-author and I reflected on a future state of project management *"Shape the future. The possibilities open to you, to all of us as the custodians of the project profession, are pretty limitless. The opportunity to shape what happens next is exciting. But shaping the future will mean losing or letting go of some of the past.*[1]

We know that to deliver change we'll need to learn together, to learn from each other but, in many cases, our project constructs aren't driven by learning and reflective practice, they aren't designed to thrive on ideas, to be driven by curiosity, to act as catalysts for innovation. They are, instead, designed to do what we used to understand as project delivery. Project delivery that is before contexts became ambiguous or got volatile, before life became more disposable to the point that no matter what you give someone will ask for more.

Project management wasn't ever developed to constrain, control or limit, and it has already started to fall short in terms of what we need now and what we'll need next.

We are not saying this to rubbish those who have gone before us, or those that currently stand with us. We say it to help open eyes that the project management profession needs to be more, it needs to be different, and it needs to keep growing and learning.

This is the part you play. The future is yours. What you choose to do next will determine whether there is a better way.

You need somewhere to channel your ideas, to set sights on ambition and respond to challenges.

What happens next needs your skills and experience (or lack of, more importantly) so that you can start to solve your own problems, moving the profession forward, just keep moving on and up."

This was stated in "Project Management: It's all ********!",[2] a slightly irreverent consideration of project management today and, potentially, tomorrow.

DOI: 10.4324/9781003346470-50

I hope that in this book you have seen an image of a PMO that represents what we wrote a few years back, and that we are challenging the future, that we are open to exciting external influences, and we are fostering a community for project managers to thrive and become the best that they can be.

Keeping it simple but keeping it effective.

Notes

1 We are talking "Unlearning" here, as discussed earlier, with more thoughts in the Appendices.
2 Project Management: It's All Bollocks! The Complete Exposure of the World of, and the Value of, Project Management – Palmer-Trew, S and Taylor, P – Publisher Routledge 2019.

Part 9

Best PMO tips

A summary of the best PMO tips is covered in this book for easy reference, together with just a few more words of "wisdom".

Consider "sanity checking" your own PMO (or your PMO roadmap) against this list as some form of objective evaluation.

DOI: 10.4324/9781003346470-51

Best PMO tips

TIP #1	Consider getting some objective help in developing your own PMO roadmap, whether you are starting from nothing or, more likely, you are doing a rebuild – even if this is just another PMO leader from outside your organisation, it doesn't have to be a professional PMO consultant.
TIP #2	The best PMOs balance all of what they need to do in order to achieve the most effective development of capability, representation of capability, and sharing of capability and achievement – your "balance" will be unique to your organisation so take reference points from other PMOs (and your past PMOs') but make it your own.
TIP #3	The best PMOs will consider investing in the "Outcomes" focus for their future PMO model – bridging that potential gap between the PMO "engine" and the stakeholders within the business.
TIP #4	The best PMOs have the very best people as part of the PMO team and ensure that they have a mix of project/ programme knowledge and remain connected to the "real world" of the project managers – evaluate those on your existing team and consider who from your network might add value to your PMO, and leave an open-door for people to ask to be involved.
TIP #5	The best PMOs have a whole lot of fun as part of their work, all of which contributes to a positive mindset and a good work/life balance – personally, I love this aspect since I have always believed in fun (the right level and right sort of fun) so reach out to your PMO team for their ideas of what might be good fun.

DOI: 10.4324/9781003346470-52

TIP #6   The best PMOs will also consider working smarter and not harder as their core mantra – challenging all that they do and questioning, "Is this the most efficient way to work?"

TIP #7   The best PMOs have a strong and recognisable identity within their own organisation – test the waters, randomly speak to people in your organisation to see if they a) have heard of the PMO and b) know what it does.

TIP #8   The best PMOs have the most experienced PMs in place and have a program underway to recruit the best PMs, develop their existing PMs into the best, and maintain this level of quality and experience – at the end of the day, it is all about people so make your PMO hyper-attractive to the best talent out there.

TIP #9   The best PMOs are the custodians of a dynamic framework of method to assist PMs in the delivery of projects. This includes not only process but also templates, guidance, and knowledge sharing – simple test, does your method work equally for the most inexperienced PMs and the most experienced PMs? Or does it hamper the best and/or not support the new?

TIP #10   The best PMOs sponsor training and facilitate communities of practice to promote PM best practices in their organisations. Such communities of practice provide PMs with a forum to share their knowledge and share experiences and have a common, open, progressive career path opportunity – as we have done, a short survey to validate engagement and those 1:1s will give you real insight.

TIP #11   The best PMOs have consistent, repeatable PM practices across the enterprise. All projects are held to the same standards and requirements for success. They have also eliminated redundant, bureaucratic PM practices that have slowed down projects – test this by running virtual projects and seeing where the overloads might be, have a feedback process for real project commentary, and keep evolving the means to project delivery.

TIP #12   The best PMOs ensure that quality assurance actually delivers quality – just make sure, really make sure, it is appropriate to need and hit it with the KISS test on a regular basis.

TIP #13   The best PMOs are not stuck in their ways but are proactively exploring opportunities to be better and to serve their sponsoring organisation and their project community better – consider how far ahead you are thinking for your PMO, how far into the realistic future your roadmap is, and how open you are to new opportunities.

TIP #14   The best PMOs always measure progress and adjust as needed based on feedback – make those measurements valid and valuable and nothing else.

TIP #15   The best PMOs are never quiet about achievements and successes but shout out loud to thank contributors and showcase progress – be bold, be brave, and be always talking about the PMO successes.

TIP #16   The best PMOs adapt to the enterprise's strategic expectations and know how to operate effectively within the corporate structure and culture. And they are not rigid in their own structure and focus in order to adapt and adopt quickly – close the loop on a quarterly basis with business strategic objectives, make sure you are aligned, and get as close as you can to the creators of strategy.

TIP #17   The best PMOs have a good understanding of the principles of change and the best approaches to change management (and adoption) – and adopting a formalised change model or process can aid this enormously, but at the very least the members of the PMO should have a strong appreciation of change management.

TIP #18   The best PMOs have clear visibility into the progress and cost of all projects. They also know exactly how resources are being used. They openly share this information with all the appropriate stakeholders throughout the enterprise – be honest and clear (and do something about the insights you might receive).

TIP #19   The best PMOs (I strongly believe) take time to reflect on the past as well as the future – just ensure that you set aside time, with your PMO team, to look back, celebrate, learn lessons, adjust, look ahead, and plan for the future.

TIP #20   The best PMOs keep it simple but effective! And probably buy a copy of this book for all their team members ... maybe even get the author in to speak at their next event.

Appendices

Meeting the author and major contributors, as well as exploring some more information about some of the topics referenced in the book. Plus, a final message for the future.

Author

Peter Taylor

Keynote speaker and coach, Peter is the author of the number-one best-selling project management book "The Lazy Project Manager", along with many other books[1] on project management, PMO development, executive sponsorship, transformation leadership, and speaking skills (Figure 10.1).

Peter has built and led some of the largest PMOs in the world with organisations such as Siemens, IBM/Cognos, Kronos/UKG, and now Ceridian, where he is the VP of Global PMO.

He has also delivered over 500 lectures around the world in 26 countries and has been described as "perhaps the most entertaining and inspiring speaker in the project management world today".[2]

You can discover more about Peter through his website, www.thelazyprojectmanager.com, and also through his podcast, The Squid of Despair,[3] at www.squidofdespair.com

Other Routledge publications

- Project Management: It's All Bollocks
- Make Your Business Agile: A Roadmap for Transforming Your Management and Adapting to the "New Normal"
- AI and the Project Manager: How the Rise of Artificial Intelligence Will Change Your World
- Team Analytics: The future of high-performance teams and project success

Figure 10.1 Peter Taylor.

Major contributors

Bill Snow

Director Global PMO – Leading the "Projects" team

Highly engaged, results oriented, and people-first operations and business leader. Successfully started up new organisations, transformed businesses, and driven operational excellence. Squarely focused on the servant-leader philosophy and the service profit chain. Thrives working within collaborative and fast paced cultures, where bold decision making is encouraged.

Expertise and passion around:

- Talent development and employee engagement
- Business process improvement

- Change management
- Programme management
- Software as a service (SaaS)

Kim DiMauro

Director Global PMO – Leading the "Methods" team

With over a decade of project management experience leading complex projects, I've advanced my career by joining the Project Management Office, whereby implementing process improvements to increase customer satisfaction and boost employee performance and successfully driving, collaborating, and navigating project managers through challenging situations ranging from risk mitigation, budget constraints, and managing resources.

I am a PMO leader with a unique knack for streamlining processes and building relationships across all verticals to create impactful solutions. Over the past couple of years, I have grown as a PMO leader and sharpened my leadership, communication, and organisation skills, which has allowed me to develop practical project health diagnostic tools to obtain key insights into project failures and factors for success.

The highlight of my job is when I get to work with new project managers and provide guidance to help with successful project implementation. I've been fortunate to have the opportunity to enhance my problem solving and creative skills at Aptos, which has enabled me to better support my internal customers who in turn can provide better support to our external clients. Each year I set career goals; this year I'm focused on up leveling my knowledge on agile methodologies and conflict resolution skills.

Thomas Neumeier

Director Global PMO – Leading the "Outcomes" team

PMP, IPMA B

International programme/project management experience in EMEA, US, and APAC. He is very experienced at developing and maintaining integration of complex programme and project management processes and systems, including knowledge management. A highly passionate, motivated, and experienced individual with proven programme management success as an executive project manager.

David Ayling-Smith

David is an experienced executive with 25 years experience of selling, delivering, and managing consultancy, support, and maintenance operations in the enterprise software industry.

David has held various leadership positions with a track record of developing strategies that deliver profitable revenue growth and scalable business improvement. Having particular experience with acquisitions and business transformations, David (aka DAS) is a strong communicator and people focused executive with a history of building successful teams and coaching individual key players.

Figure 10.2 Festive team leaders.

Left to right: Peter, Thomas, Kim, and Bill in Boston 2023.

Wall of fame

My full Global PMO team where our motto is "Together Everyone Achieves More" is:

- Bill Snow (USA)
  - Kathleen Rousseau (USA)
  - Brandy Kirsch (USA)
  - Ahmad Jamal (Canada)

Methods

- Kim DiMauro (USA)
  - Lauren Shankle (USA)
  - Erika Williams (USA)
  - Sally MacQueen (USA)

Outcomes

- Thomas Neumeier (Germany)
  - Jost Lange (Germany)
  - Jason LaPrade (Australia)
  - Anand Surendran (Australia)
  - Matt Parker (USA)
  - Mitch Rajmoolie (Canada)

Some team thoughts

Lauren – *"Being part of an active and successful GPMO is to know you are bettering work life for hundreds of others. It is energising, inspiring, and uplifting"*.

Matt – *"People often wonder what does a PMO do? The answer varies based on who you ask. Once I joined the Global PMO at Ceridian, I found clarity – Projects (education and training), Methods (methodology refinements and best practices), and Outcomes (regional field support and connections). Our PMO does not arbitrarily make changes, we strategically develop our road map based on these three fundamentals and the deep connections with the field which in turn drives successful projects and happy clients. Seeing this in action is powerful and shows a PMO can make a highly impactful difference"*.

Mitch – *"Working with the Ceridian Global PMO has been the highlight of my career. It has given me the opportunity to work with people from diverse backgrounds and cultures; enabling me to learn new perspectives and ways of thinking"*

Bill – *"Building a PMO can be an exciting an experience – aligning to organizational vision, creating the PMO mission and roadmap, designing teams, creating an ecosystem of great partners, and turning concepts into real value, all for the benefit of our people, business, and customers, is very rewarding!"*

Ceridian HCM Inc

Ceridian is a global human capital management (HCM) software company. Dayforce, our flagship cloud HCM platform, provides human resources, payroll, benefits, workforce management, and talent management capabilities in a single solution. Our platform helps you manage the entire employee life cycle, from recruiting and onboarding to paying people and developing their careers. Ceridian provides solutions for organisations of all sizes, from small businesses to global organisations.

Ceridian's brand promise is "Makes Work Life Better". It embodies who we are, what we believe, and what we stand for. We deliver on that promise by aiming to improve the work lives of our customers, their employees, and our own teams at Ceridian every day and with everything we do.

https://www.ceridian.com

Our stellar speakers

It was noted earlier in the "Project Community" chapter that we ran "Inspire" sessions with an amazing range of external speakers, to bring some objective insight and learning.

If your PMO, your organisation, is looking for some top speakers,[4] then you might consider these folks – they are all highly recommended:

Speaker	Location	Topic(s)
Lee Lambert https://www.linkedin.com/in/lelambert/	USA	PMP and the value of this investment
Fatimah Abbouchi https://www.linkedin.com/in/fatimahabbouchi/	Australia	Delivery done differently, putting the agile into governance, not the governance into agile
Rick Morris https://www.linkedin.com/in/rickamorris/	USA	Ethical influence, understanding the power of persuasion and how that impacts projects and teams
Marissa Silva https://www.linkedin.com/in/theluckypm/	Portugal	Once upon a time, there was a project.
Joe Pusz https://www.linkedin.com/in/joepusz/	USA	PM and making Pizza

(Continued)

Speaker	Location	Topic(s)
Gary Nelson https://www.linkedin.com/in/ garymnelsonpmp/	New Zealand	"Never too young to be a project manager" – the importance of early education/the universality of PM principles and how they apply to much of daily and working life
Alfonso Bucero https://www.linkedin.com/in/alfonso-bucero- msc-ace-pmp-rmp-pfmp-ipmo-e-pmi- fellow-97348a/	Spain	Passion, persistence, and patience to manage projects
Thomas Walenta https://www.linkedin.com/in/thwalenta/	Germany	Leadership and individual resilience
Geoff Crane https://www.linkedin.com/in/ageoffreycrane/	Canada	Personal Intelligence construct and how it's different from EI
Susie Palmer-Trew https://www.linkedin.com/in/susie-palmer- trew-13108646/	UK	Really Smashing Change Leadership

Community recording guide

Recording options

As noted in the "Project Community" chapter, not all communication sessions need to be live sessions. Sometimes, to provide timely communications to the organisation, it may be best to conduct a closed recording session by working directly with your contributor to create the video. If you're unable to coordinate such a closed recording session with the guest speakers, here are a couple of options you can try:

- Option 1 – Speaker creates an mp4 you'll add to your deck:
  - Speaker recording: Plan to use your presentation, and instead of adding your speaker's slides into your deck, have them create a recording and provide you the mp4
  - The Plan:
    - Let your speaker know your plan to open and close the video as usual (with or without music), and that you'll introduce them before triggering their video

- For this to appear natural, recommend that they begin their recording by saying thanks to you and that they're happy to be there
- Then they can launch into their topic and close their recording by saying thanks again to you, for the opportunity to present their topic
- Finally, once you have their mp4, you can drop their video into your deck, and use your recording application of choice, be sure to use the feature to "share audio", and record the entire presentation start to finish including your opening, your speaker's mp4 content, and your closing wrap up in which you'll thank the speaker and your audience for their support, before playing your closing music and ending the video recording

- Making It Happen:

  - Have your presentation deck template prepared
  - Receive your speaker's mp4
  - Use PowerPoint's features:

    - Include a blank slide after your title and introduction slides.
    - Select Insert > Video > This Device > Locate and select the mp4 from your desktop.
    - Once everything's ready, select "Slideshow" mode and check to confirm the video will trigger and that audio will play.
    - Adjust the video's sound directly in the PowerPoint slide, if needed.
    - Next, it's show time! Use your preferred method to record. Kick things off with your title slide and introduction. Trigger the video to play. Once completed, wrap up the recording with thanks, and music, if applicable, and close your recording.
    - Double-check your finished video to ensure everything is in good order.
    - Consider using a tool, like Camtasia, to polish the final video, if needed.

- Option 2 – Speaker adds audio to each slide and provides you with their deck:

  - Speaker recording: This option is particularly good for those with a bit of stage fright, or those that don't like to be recorded. Have your speaker prepare their own slides, adding audio to each of the slides in their deck. This gives speakers the chance to make very small, scripted recordings, and they'll be in control if they'd like to keep or re-record the content on each slide.

- The Plan:

  - Like the prior option, be sure your speaker knows your plan to open and close the video as usual (with or without music), and that you'll introduce them before triggering their video
  - For this to appear natural, recommend that they begin their recording by saying thanks to you and that they're happy to be there
  - Then they can use the steps outlined below to add audio to each slide of their deck
  - For their last slide, ask the speaker to close their recording by saying thanks again to you, for the opportunity to present their topic
  - For this option, the speaker will not need to create an mp4. Instead, just have them provide you with their presentation deck to you
  - Double-check to ensure everything is in good order

    - Note: If you can see audio is embedded, but can't hear it, the guidance below shows how you may "unmute" the audio on that slide, and/or adjust the volume

  - Use your recording application of choice, make sure you're sharing audio, record your opening, and continue to advance slide by slide through the speaker's content with audio all the way through your closing wrap-up in which you'll thank the speaker and your audience for their support, before playing your closing music, and ending the video recording

- Making It Happen:

  - Have your presentation deck template prepared
  - Share this guidance with your guest speaker:

    - Wear a headset for the best audio quality
    - Use PowerPoint's features to:

      - Select Insert > Audio > Record audio

    - From the "Record Sound" pop-up, select the radial button and begin speaking
    - Once finished, you have a few options:

      - Click "OK" to save the recording as is
      - Click "Cancel" to delete the recording

- Additional options include:
  - Clicking the square shape to stop the recording and the triangular shape to listen to the recording
  - Once you listen to the recording, you can click "OK" to save the recording or click the radial button to re-record
- Troubleshooting: Can't hear audio, or need to adjust the volume?
  - Click the grey audio icon on screen to expand options
  - If you can see by the grey bar at the bottom that the audio is progressing, but cannot hear the sound, click the tiny audio icon image in the lower-right corner that's a "mute button" to unmute the sound
  - If you need to adjust the volume, use the volume slider above the mute button as needed

Other reference books

Leading successful PMOs[5]

This is a comprehensive guidebook for establishing and managing effective Project Management Offices (PMOs). This book offers practical advice, real-world examples, and case studies to help readers build the best PMO for their organisation.

One of the strengths of the book is its clear and straightforward writing style. Taylor does an excellent job of explaining complex concepts and technical jargon in a way that is easy to understand. The book is also well-structured and organised, making it easy for readers to follow along and find the information they need.

Another highlight of the book is its focus on the importance of executive support for successful PMOs. Taylor emphasises that without support from senior leadership, PMOs are unlikely to succeed. He also provides practical advice for gaining executive buy-in and building a business case for a PMO.

Overall, "Leading Successful PMOs: How to Build the Best Project Management Office for Your Business" is a valuable resource for anyone involved in project management or responsible for establishing and managing a PMO. It provides a wealth of practical advice, best practices, and real-world examples that can help organisations build effective PMOs that deliver value to the business.

Delivering successful PMOs[6]

This is perfect for project management professionals, managers, and executives who want to create or improve a PMO. In this book, Taylor and his co-author provide practical advice and insights based on his extensive experience in building and managing PMOs.

The book is well structured, with clear explanations of the different types of PMOs and their respective functions. Taylor provides a framework for establishing a PMO, including how to identify the right type of PMO for your organisation, how to staff and manage it, and how to measure its success.

One of the strengths of the book is its emphasis on the importance of organisational buy-in and stakeholder engagement. Taylor stresses the need to align the PMO with the strategic goals of the organisation and to involve key stakeholders in the design and implementation of the PMO.

Throughout the book, Taylor includes practical tips and case studies to illustrate his points. He also provides a helpful list of "best PMO tips" at the end of each chapter, which summarises the key takeaways.

Overall, "Delivering Successful PMOs: How to Design and Deliver the Best Project Management Office" provides a comprehensive overview of the key considerations and best practices and is written in an accessible and engaging style.

Other reference material

The "right" model

This is a "Keep It Simple" model for describing a typical PMO model and scope.

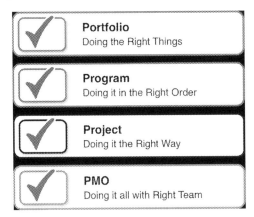

Figure 10.3 The Taylor "KISS" PMO model.

Simply put, something I try to do in all matters, the model proposes that managing a project is really just about doing it the "right way" – that is, delivering the project in the most appropriate and effective manner to bring about the best business outcome.

A programme (programme) is just about doing the right projects in the right order.

A portfolio is about having the right "things" happening to bring about the expected business outcomes, so the right projects and programmes.

And a PMO, a great PMO, is successful because it has the "right team" in place.

Traditional "PMO" definitions

For reference only – make yours what it needs to be! And call it whatever you bloody well want to, it really doesn't matter – it's the reputation that counts!

- PSO – Project Support Office

  - Some organisations employ this term to describe the provision of administrative support rather than project management assistance.

- PMO – Project Management Office

  - The PMO assists in the delivery of projects, offering varying levels of management and support depending on the maturity of the PMO and the organisation.

- PgMO – Programme Management Office

  - Typically, the PgMO oversees multiple projects that constitute a programme. This often involves PMOs reporting to the PgMO.

- PPMO – Project Portfolio Management Office

  - The PPMO constructs a comprehensive view of projects and programmes within a defined portfolio. While they may facilitate delivery, most organisations use them to make decisions about which projects to execute and when.

- EPMO – Enterprise Project/Programme/Portfolio Office

  - This typically serves as the overarching PMO responsible for standards, strategy, direction, and more. It connects to the PMO network to ensure that all projects align with the strategy and adhere to established standards. It is particularly popular among organisations that aim to translate their strategies into change programmes and execute them.

Be a great PMO leader

The key skills seem to be that a PMO leader needs to have a passion for projects (some of the PMO failures I have come across can be attributed to the fact that the PMO head had no project background experience).

Secondly, they need to be great communicators and strong negotiators to help the PMO find its place inside the organisation and explain the value.

They also needed to be enthusiastic about leading change.

And finally, they must not be afraid to tailor their PMO to the unique model that suits the business.

The lazy winner

The book "The Lazy Winner"[7] provides five tips for achieving more with less effort and improving work-life balance.

Do you want to do more with less effort? Do you desperately want a better work-life balance? Then you need to start by asking yourself some fundamental questions when the next job, piece of work, task, or request comes your way.

1 Question the necessity and worth of each task: Instead of blindly accepting work, ask yourself if it's necessary and worth doing. Don't feel obligated to do something just because others are doing it.
2 Focus on high-impact tasks: Apply the 80/20 rule and prioritise tasks that deliver the most value. Invest your time in activities that yield the highest return on your personal investment.
3 Delegate when possible: Determine if someone else is better qualified to handle a task and let them help you. Allocate work to the most suitable person to benefit everyone involved.
4 Find the shortest path to success: Avoid unnecessary complexity and streamline tasks. Simplify, eliminate, or automate steps to accomplish the goal efficiently. Consider if a task can be delayed without negatively affecting others.
5 Seek long-term value: Explore ways to make tasks reusable, scalable, and of greater value in the future. Look for opportunities to automate, scale, or create wider applications for your work.

By following these tips and considering the overall impact and value of tasks, you can optimise your personal return on investment and achieve more while exerting less effort.

Unlearning

In the 21st century, unlearning has emerged as a vital skill. With age comes wisdom, and I have come to realise that throughout my life and career, there have been instances where I had to discard or "unlearn" what I had previously learned.

Contrary to the belief that learning is a cumulative process, it is actually a series of transformative shifts in understanding and knowledge. As Albert Einstein aptly stated, "The world as we have created it is a process of our thinking. It cannot be changed without changing our thinking".

In the face of rapid global changes, it is crucial for us to continuously learn new things and embrace a mindset of perpetual learning to remain relevant. However, the true challenge lies not in learning new information but in unlearning our past knowledge and habits.

The inspiration behind Barry's book, "Unlearn", stemmed from observing highly intelligent individuals who, despite their aptitude for learning, struggled to let go of their past beliefs and behaviours, especially if they had contributed to their previous success.

As humans, we often operate on autopilot, driven by ingrained patterns and subconscious responses. Unlearning the past is not an easy task due to the nature of our brains, which are wired to seek familiar patterns. Our brains, with billions of neurons and trillions of connections, exhibit two dominant drivers: Self-preservation and efficiency. Being the command centre of our bodily functions, the brain aims to conserve energy by auto-mating processes. Neuroscience suggests that 95% to 97% of our brain's functioning is driven by automatic responses rooted in the subconscious mind. This realisation highlights the challenge of recognising when and how to break free from deep-seated mindsets, habits, and responses.

When we need to learn a new approach that contradicts our established beliefs, it becomes essential to first let go of the old ways. Otherwise, learning the new method becomes incredibly difficult as it conflicts with our existing framework.

Unfortunately, we all suffer from a natural bias known as the "sunk cost" fallacy when it comes to our knowledge and skills. If we have invested a considerable amount of time in acquiring expertise, it becomes challenging to accept that it may have become obsolete due to changing circumstances. Moreover, learning is not always a gradual and cumulative process; at times, we must completely abandon old, obsolete thinking and approaches to make room for new and improved methods.

So, what exactly is unlearning?

According to Barry O'Reilly's bestselling book, "Unlearn: Let Go of Past Success to Achieve Extraordinary Results",[8] unlearning entails con-sciously releasing and reframing once-useful mindsets and acquired behaviours that were effective in the past but now hinder our success. Unlearning does not entail forgetting or discarding knowledge and experience; rather, it involves actively letting go of outdated information and engaging in acquiring new knowledge to inform effective decision making and action.

Therefore, to ensure our continued relevance, we must be open and willing to unlearn outdated mindsets and behaviours if our reality undergoes significant changes. Unlearning is not about abandoning everything we know; it's about being prepared to discard obsolete information when necessary.

It is worth noting that we already undergo changes and adaptations as humans, often unconsciously and out of necessity. The Events of 2020 demonstrated this truth. However, what sets the present apart is that we now live in a world characterised by constant and increasingly rapid change.

If we do not make unlearning a frequent, consistent, and deliberate practice, we risk becoming obsolete, and this obsolescence can occur swiftly. What has worked well in the past may not work well in the future.

Team building

There is a real power to teams, in that they are more likely to achieve larger goals than individuals alone, and they are more likely to be innovative, more productive, and more adaptive as changes in the marketplace or workplace occur.

The advantages of teams over individual workers are very clear. The successes that I have had over the years in project delivery and PMO leadership could not have happened without performing teams. None of this is a solo effort.

There are five principles for enhancing project team performance.

#1: Acceptance that it's time for a new approach
One cannot go about making a change effectively if they don't believe that change is necessary. The first step to progress is to accept the necessity for progress to be made. No matter what the past or current success levels are, there is always room for improvement.

You need to really understand the importance of a people-centric approach. After all, these are the people who work on the front lines every day; those who can make an immediate difference to an organisation's performance. They can do this by starting with team performance and team well-being to ultimately achieve team success.

#2: Empowering managers has never been more important
The world is changing at an increasingly fast pace. Because the pandemic has accelerated the transition to remote work and in-person interactions

(both formal and informal) have fallen by the wayside, it has never been more important for managers to find ways to keep their fingers on the pulse regarding their teams' performance and engagement. Working virtually will inevitably make issues harder to notice and will increase the likelihood of problems sneaking up on unsuspecting managers.

As the rate of change accelerates, organisations must become more agile. Those who can adapt rapidly will survive; those who can't, won't. To make the right decisions swiftly, managers need data delivered in real-time. They need leading data points on team performance that can help them tackle issues as they arise – not lagging data points that highlight issues after they've come to the fore – measuring performance not merely to evaluate team members but to empower them.

Without timely insights, managers are essentially left flying blind, unable to understand the human factors that impact team dynamics, performance, and engagement.

#3: This change is critical for all project-based teams
As we've stated, the need for real-time performance data is particularly pressing for managers of project-based teams. The way these cross-functional teams operate and the way they are assembled and disassembled means that they face greater uncertainty and urgency than functional teams do. As a result, a great deal of agility is required from project-based teams.

To succeed consistently in our projects, we must give managers and teams the insights they need to solve problems collectively, before they escalate, and to make more informed decisions. This need will only become more apparent with the ongoing shift toward project teams and away from traditional teams in our world of work.

#4: Identifying areas for improvement
Most organisations are doing some form of team feedback, retrospection, or assessment already, albeit ineffectively. It is generally too infrequent, not mandated, systematic, or automated – but it's a great way to start by identifying the closest thing to what you already have and expand on that as a low-hanging fruit.

Since retrospectives are unfortunately mostly focused on work and not the team, they're not always inclusive – often the same team members contribute their opinions, with the quieter bunch staying silent.

We've seen many teams start improving their retro practice by implementing a data or analytics tool and then rocking up to a retro equipped with insights and ready to act together as a team – often already having asynchronously found solutions. This also allows all team members to participate and provide suggestions without feeling any form of trepidation or fear of scrutinisation.

#5: Start today, not tomorrow

I truly hope that, through reading this article, and other research and experience, you will reach the (inevitable) conclusion that this change is coming.

The call to action is simply, start today, don't leave it to see what happens to other organisations or what happens to other teams. Instead, take the lead, take control, take the advantage.

As noted in principle #4, you are, most likely already doing this in some way and this can form the platform for progression to a much more productive world of project team performance, the world of team analytics.

First published https://www.capterra.com/resources/project-management-team-performance/ 2023.

Four genius ideas

… from the team at Amazon (well, I think they are pretty neat – have a read, have a think, and see if may might work for your team)

1 The two-Pizza Team
2 The institutional Yes
3 The Future Press Release
4 The Empty Chair

So, picture this: Jeff Bezos, the mastermind behind Amazon, came up with a concept called the "two-pizza team". Now, I know what you're thinking: "Two pizzas? What's the big deal"? Well, it turns out that at Amazon, they believe a team should be small enough to be fed with just two pizzas. That's right, two pizzas! I mean, who knew pizza could be the secret ingredient to success?[9]

The idea behind the two-pizza team is to create teams that are as nimble as a squirrel on caffeine. They want these teams to be small, agile, and so autonomous that they could probably run a marathon without breaking a sweat. Because let's be honest, when you're chomping down on a slice of pizza, you don't have time to deal with bureaucratic nonsense.

By keeping the team size small, Amazon aims to avoid those pesky bureaucratic hurdles that slow down progress. They want their teams to communicate faster than the speed of light, make decisions at the drop of a hat, and take ownership of their work like a boss. It's like a game of hot potato, but instead of a potato, it's a responsibility, and instead of burning your hands, you're just making good things happen.

[Peter: My GPMO fits this model]

But wait, there's more! Amazon has a concept called the "Institutional Yes". Now, in most companies, the default response to an idea is a big fat "No". It's like they have a committee whose sole purpose is to shoot down dreams faster than a malfunctioning slingshot. But not at Amazon, oh no! At Amazon, the default response is a resounding "Yes!" If someone wants to say no, they have to write a full-page essay on why the idea sucks. Talk about motivation to get on board!

[Peter: Challenging and perhaps not an approach that you take for every idea but for the biggies, then why not give it a go?]

And hold onto your hats, because we're about to take a ride on the "Future Press Release" train. Imagine this: Before a project even starts, the team creates a document that's basically a sneak peek into the future. They write up a press release announcing the glorious outcome they hope to achieve when their project goes live. It's like they're time travellers, except they're not really travelling through time, they're just really good at imagining stuff.

[Peter: It works!]

But hey, it's not all fun and games at Amazon. They take their customers seriously, like really seriously. In important meetings, they bring in an empty chair and pretend that a customer is sitting there, judging their every move. It's like the ghost of customers past, present, and future haunting their decision-making process. They want to make sure those customer needs and wants are met, even if it means summoning a chair from the void. That's dedication!

[Peter: It is certainly all too easy to forget your true customers (internal or external) and get all subjective in your own PMO world – beware!]

The Marshmallow challenge

The Marshmallow Challenge is a team-building exercise that involves building a tall structure using only spaghetti sticks, tape, string, and a marshmallow. The goal is to construct the tallest freestanding tower that can support the weight of the marshmallow on top. Here's a simple guide to the Marshmallow Challenge:

Step 1: Gather the materials:

- Spaghetti sticks (dry, uncooked) – 10 per team
- Marshmallows (one per team)
- Tape (masking tape) – 1 yard or metre per team
- String or yarn – 1 yard or metre per team
- Scissors – 1 per team

Step 2: Form teams by dividing participants into small teams of 3–5 people. Each team will work together to build their tower.

Step 3: Clearly explain the objective of the challenge to all teams. The goal is to construct the tallest tower possible using the given materials, and the marshmallow must be placed on top.

Step 4: Set the rules:

- The tower must be freestanding, meaning it cannot be attached to any other objects or structures.
- The marshmallow must be placed on top of the tower and should not be impaled on any of the spaghetti sticks.
- All the materials provided can be used, but nothing else (no additional tools or objects).
- The tallest tower that can support the marshmallow's weight wins.

Step 5: Go! Start the timer: 18 minutes creates a sense of urgency. Play some music with increasing tempo.

Step 6: Measure up: When the time is up, measure the height of each team's tower from the base to the highest point. Make sure the marshmallow is still on top and not touching any other surface. Any nonstanding tower is disqualified. Award a prize to the winning team.

Step 7: Debrief and discuss: Once all the towers have been measured, gather the teams together for a debriefing session. Discuss the different strategies, challenges faced, and lessons learned during the activity. Encourage participants to reflect on the importance of teamwork, creativity, and problem-solving.

Play this great video with Marshmallow Challenge insights https://www.ted.com/talks/tom_wujec_build_a_tower_build_a_team Tom Wujec[10] "Build a tower, build a team".

The Marshmallow Challenge is a fun and engaging activity that promotes collaboration, creativity, and innovation. It's a great way to foster teamwork and explore the dynamics of group problem solving.

Maslow's hierarchy of needs

Maslow's Hierarchy of Needs is a psychological theory proposed by Abraham Maslow in 1943, which suggests that human beings have a set of hierarchical needs that must be fulfilled in a specific order to achieve self-actualisation and personal growth. The hierarchy is typically depicted as a pyramid with five levels, representing different categories of needs:

- Physiological Needs: This is the base level of the hierarchy and encompasses the most fundamental needs necessary for survival, such as air, water, food, shelter, sleep, and clothing. These needs must be satisfied first, as they are essential for sustaining life.
- Safety Needs: Once physiological needs are met, individuals seek safety and security. This includes protection from physical harm, a stable and secure environment, financial security, health, and personal well-being. Safety needs also extend to emotional security and the absence of fear or threat.
- Love and Belongingness Needs: At this level, individuals have a need for social connection, love, and a sense of belonging. This includes forming meaningful relationships, experiencing intimacy, and being part of a family, friendship, or community. Meeting these needs involves giving and receiving affection, acceptance, and companionship.
- Esteem Needs: The next level involves the desire for self-esteem and the recognition and respect of others. Esteem needs can be divided into two categories: Internal (self-esteem, self-confidence, self-worth) and external (recognition, status, reputation). Fulfilling these needs helps individuals develop a sense of competence, achievement, and confidence.
- Self-Actualisation: This is the highest level of the hierarchy and represents the need for personal growth, fulfilment, and self-actualisation. Self-actualisation involves reaching one's full potential, pursuing individual goals, and finding meaning and purpose in life. It includes aspects such as creativity, problem solving, personal development, and the desire for personal growth.

According to Maslow, individuals progress through the hierarchy in a sequential manner. The lower-level needs must be largely satisfied before higher-level needs become motivational. However, it is important to note that the hierarchy is not rigid, and individuals may prioritise different needs based on their unique circumstances, culture, and personal values.

Maslow's Hierarchy of Needs provides a framework for understanding human motivation and the factors that drive individuals to seek personal fulfilment and self-actualisation. It has been influential in fields such as psychology, education, and management, shaping our understanding of human behaviour and well-being.

Kübler-Ross's five stages of grief

Elisabeth Kübler-Ross is known for her work on the five stages of grief. These stages are often experienced by individuals facing a terminal illness or grieving the loss of a loved one. Here are the five stages:

1 Denial: The first stage involves a sense of disbelief or denial in response to the loss or impending death. It serves as a defence mechanism to protect individuals from overwhelming emotions. People may struggle to accept the reality of the situation and may reject or minimise the loss.

2 Anger: In the second stage, individuals may experience anger and frustration. They may feel resentful and place blame on others or themselves for the situation. This anger can be directed at healthcare professionals, the deceased, or even a higher power.

3 Bargaining: The bargaining stage involves attempting to negotiate or make deals to change the outcome or delay the loss. People may make promises, pray, or seek alternative solutions in an effort to regain control or postpone the grief-inducing event.

4 Depression: The fourth stage is characterised by a deep sense of sadness and profound loss. Individuals may experience feelings of emptiness, loneliness, and despair. They may withdraw from others, lose interest in activities, and experience changes in appetite and sleep patterns.

5 Acceptance: The final stage is acceptance, where individuals come to terms with the loss or impending death. It does not mean that they necessarily feel okay or happy about the situation, but rather that they have reached a level of understanding and begin to find ways to move forward with their lives.

It's important to note that not everyone experiences these stages in the same order or with the same intensity. The stages of grief can be fluid, and individuals may move back and forth between them. Additionally, these stages are not intended to be prescriptive or all-encompassing, as grief is a highly individual and personal experience.

Elisabeth Kübler-Ross's model has had a significant impact on our understanding of grief and has been widely used to help individuals and their loved ones navigate the emotional challenges associated with loss and terminal illness.

Theoretical learning models

Theoretical learning models provide a framework for understanding how individuals acquire knowledge, skills, and understanding. One such prominent theoretical learning model is the cognitive-behavioural model, which focuses on the cognitive processes involved in learning and how they interact with behaviour. The model emphasises the role of internal mental processes, such as perception, memory, attention, and problem-solving, in shaping behaviour and learning outcomes.

The cognitive-behavioural model can be broken down into several key components:

- Input: In this stage, individuals receive information from their environment through sensory perception. This input can be in the form of visual, auditory, or tactile stimuli.
- Encoding: The information from the environment is processed and encoded in a way that can be understood and stored in memory. This involves organising and transforming the incoming information based on existing knowledge and schemas.
- Storage: Once encoded, the information is stored in memory. The cognitive-behavioural model recognises the importance of long-term memory, where information is retained for an extended period and can be retrieved later.
- Retrieval: When needed, the information stored in memory is retrieved and brought into consciousness for use. This stage involves accessing and recalling the relevant knowledge or skills to guide behaviour or solve problems.
- Reinforcement: The cognitive-behavioural model emphasises the role of reinforcement in learning. Reinforcement can be positive (rewarding) or negative (removing something unpleasant), and it strengthens the association between a behaviour and its consequences. Reinforcement increases the likelihood of the behaviour being repeated in the future.
- Feedback: Feedback is essential for learning and adjusting behaviour. It provides individuals with information about the accuracy or effectiveness of their actions and helps them adjust and improve their performance.
- Transfer: Transfer refers to the ability to apply knowledge or skills learned in one context to another context. The cognitive-behavioural model recognises that learning is not limited to specific situations and should ideally facilitate transfer to real-life situations.

This theoretical learning model highlights the importance of cognitive processes, such as perception, memory, and problem solving, in shaping behaviour and learning outcomes. It emphasises the interaction between cognition and behaviour, with the goal of understanding and explaining how individuals learn, acquire new skills, and develop a deeper understanding of the world around them.

AKDAR model

1   The AKDAR model is a framework that is commonly used in communication and conflict resolution to guide individuals through difficult conversations and help them achieve positive outcomes. AKDAR is an

acronym that stands for Acknowledge, Discuss, Assess, and Respond. Each step represents a key component of effective communication and conflict resolution.

2 Acknowledge: The first step in the AKDAR model is to acknowledge the issue or concern being raised. This involves actively listening to the other person, validating their feelings, and showing empathy towards their perspective. Acknowledging the problem helps create an atmosphere of trust and openness.

3 Discuss: In the second step, the parties involved engage in a constructive discussion about the issue at hand. This involves expressing their own viewpoints, sharing relevant information, and exploring different perspectives. It is important to maintain a respectful and non-confrontational tone during this stage to encourage open dialogue.

4 Assess: After discussing the issue, the next step is to assess the situation from various angles. This involves evaluating the potential impact and consequences of different actions or solutions. It is important to consider both short-term and long-term implications and identify any potential trade-offs or risks associated with different approaches.

5 Respond: The final step of the AKDAR model is to develop and communicate a well-thought-out response or solution. This response should consider the insights gained from the previous steps and address the concerns and interests of all parties involved. It is crucial to be clear, specific, and considerate in your response to ensure effective communication and a positive resolution.

The AKDAR model emphasises active listening, empathy, and collaboration as essential elements of effective communication and conflict resolution. By following this model, individuals can navigate difficult conversations more effectively, improve understanding, and work towards mutually satisfactory outcomes.

For more information see https://www.prosci.com/methodology/adkar

Project insanity

So why do we accept "insanity" as the path of project management?

The next time you are in a meeting just try this out.

Whether you are presenting or someone else it doesn't matter. But what happens when the inevitable happens, you go to write something on the flipchart, or the whiteboard and the pen is dry?

Have you ever (and I freely admit I am guilty) put the pen down on the rack again, picked up another one, and carried on with the point you were making? So now you have left the same dry pen for the next person. Or worse, for yourself to do the same thing again a little later in the meeting.

Did you expect the pen to magically refill itself? Of course not. That's madness!

Did you put the pen in the bin and ensure that a new one was put in its place? Or did you at least tell someone? Of course not. That's madness!

A simple lesson in lessons learned, or the process of not learning to be more precise.

Monetising the PMO

A PMO costs! It is obvious; there is the money to staff and to run the PMO together with any incurred operational expenses and systems investments and, when the PMO interfaces with other parts of the organisation – as it should, there is associated cost to that time and effort. Of course, the belief from those that sponsor a PMO is that the money and time invested will be more than saved by delivering more successful projects, and for sure, that is the primary purpose of any PMO, to deliver healthier and more successful projects appropriate to the business strategy of the organisation.

But there may come a time when there is pressure for the PMO to contribute more than that through:

- Partial cost recovery of the PMO
- Cost neutralisation of the PMO
- Profit contribution from PMO

It is important to understand any reason behind the desire to monetise the PMO. If it is driven by the PMO's success and an opportunity to expand its role, it can be a positive step. However, if it is solely to reduce costs or budget constraints, it may put the PMO under survival pressures and negatively impact its performance.

The right time to consider revenue generation is when the PMO is well-established, stable, and aligned with the organisation's strategy. Be warned against stretching the PMO too thin by simultaneously focusing on internal development and pursuing a profitable business.

To successfully monetise the PMO, keep it simple and leverage existing strengths, proven practices, and packaged services. The PMO should offer discrete, high-value services that are in demand, such as health checks or retrospectives/lessons learned.

It is crucial to avoid distractions and conflicts that may arise when serving money-generating customers while balancing the needs of budget-supporting sponsors.

In the case of external customers, the PMO's offerings should align with existing sales, marketing, and support channels and for internal services;

careful planning and communication are necessary to ensure other departments understand the value proposition and incorporate it into their budgets.

Any PMO's primary focus should be on delivering successful projects. The transition from a budget overhead to a profit contribution requires careful planning and execution.

What ChatGPT thinks

Who am I to argue with the "hot" AI tool of the moment (I will leave you to do the arguing …).

Peter Taylor is a well-known figure in the project management field. He has authored several books and is recognized as an expert in project management practices. Taylor emphasizes the importance of effective project management and offers insights into various project management methodologies.

One of the concepts Taylor has extensively discussed is the Project Management Office (PMO). A PMO is a centralized unit or department within an organization that oversees and manages projects. Its primary objective is to standardize project management practices, ensure consistency, and provide support and guidance to project teams.

Overall, Peter Taylor has made significant contributions to the project management field, and his insights on PMOs have helped organizations improve their project delivery and achieve greater success in managing projects.

ChatGPT is an AI language model developed by OpenAI.[11] It is based on the GPT-3.5 architecture and designed to generate human-like responses in natural language conversations. ChatGPT is trained on a vast amount of text data from the internet and various sources, allowing it to understand and generate text in a wide range of topics and contexts.

As a language model, ChatGPT can understand and respond to questions, engage in dialogue, provide information, and assist with various tasks. It leverages its pre-trained knowledge to generate coherent and contextually relevant responses based on the input it receives.

ChatGPT has been trained to be a versatile conversational agent, capable of engaging in conversations on diverse subjects. However, it's important to note that while ChatGPT can provide helpful information, it may not always have access to the most up to date or accurate information since its training was completed in September 2021. It's always a good idea to verify critical or time-sensitive information from reliable and current sources.

Keep on being proud of what you do

"I still want to shout out to the world about project management and tell everyone about all the great work that we do, and I want you to join me in that shouting – be loud and very proud of what you do. It is both essential and exciting. Taking project management out of the box will spread the word outside our community about what a great bunch of people we are and how project management is a valuable skill to pretty much everybody.
Be loud and be proud (or at least don't be shy)".

These were my words a long time ago in "The Lazy Project Manager and the Project from Hell",[12] and they are still true today.

This is as good a place as any to also reflect, and thank, all of those people whom I have helped become published in some way or other – I think it is over 80 people in total so far over the range of books that I have had published. Whether as a contributor or as a co-author – a big, big thank you.

Notes

1 Most of which he has cunningly managed to reference in this book (shameless promotion never hurts) – all his books except "Strategies for Project Sponsorship" (Management Concepts Press), "The 36 Stratagems" (Infinite Ideas), "The Art of Laziness" (Infinite Ideas), "The History of Laziness" (TLPM Publishing), "The Lazy Blogger" (TLPM Publishing), "The Projectless Manager" (TLPM Publishing), "Personal Productivity: Self-Assessment" (Bookboon), "Personal Productivity: Making the Change" (Bookboon), "The Extra Lazy Project Manager", and his children's book, "Dance of the Meerkats".
2 To be clear, this was not a quote from his mother.
3 With his co-host, David Ayling-Smith (Unscripted podcast musings on business life, leadership, creativity, transformation, and all the myriad of other work-life events that get in the way of a good night's sleep).
4 There is also this guy of course https://thelazyprojectmanager.com/ – just in case you are desperate.
5 Leading Successful PMOs: How to Build the Best Project Management Office for Your Business – Taylor, P – Publisher Gower 2011.
6 Delivering Successful PMOs: How to Design and Deliver the Best Project Management Office for your Business – Mead, R and Taylor, P – Publisher Gower 2015.
7 The Lazy Winner: How to Do More with Less Effort and Succeed in Your Work and Personal Life without Rushing Around Like a Headless Chicken or Putting in 100 Hour Weeks – Taylor, P – Publisher Infinite Ideas Publishing 2011.
8 Unlearn: Let Go of Past Success to Achieve Extraordinary Results – O'Reilly, B – Publisher McGraw Hill 2018.
9 For full disclosure, the author of this book does believe that pineapple has a place on a pizza ....

10 Tom Wujec is the author and editor of several books, a fellow at Autodesk, an adjunct professor at Singularity University, a multiple TED Conference speaker, and a pioneer in the emerging practice of business visualisation.

11 OpenAI is an American artificial intelligence research laboratory consisting of the non-profit OpenAI Incorporated and its for-profit subsidiary corporation OpenAI Limited Partnership. OpenAI conducts AI research with the declared intention of promoting and developing a friendly AI.

12 The Lazy Project Manager and the Project from Hell – Finer, M and Taylor, P – Publisher The Lazy Project Manager Ltd 2014 – Based on a workshop Peter has been delivering for several years, the information provided allows you and your team to undertake your own Project from Hell workshop where you analyse the problems and prepare an improvement plan that will be the basis for rescuing the project.

The last word

As I noted in the dedication, all the way back at the front of this book, "…as this is, most likely, my last book relating to the world of project management and PMOs", then I conclude with the hope and the challenge that many others will follow me, speaking, writing, and leading great PMOs (whatever that means in the future – PMOs are always changing and evolving after all).

Back in 2009, I was unpublished and had only presented in a work environment but the opportunity to write my first book was offered to me, and shortly after that (about six months), the need to then shift a pile of books with my name on them out of my garage set me on a path to speaking professionally and writing many more books, and for that, I am really grateful.

After 30 books over a 15-year period, together with 500[1] presentations in many countries, then I suspect the world has heard enough from me and it is time to give others a voice. But please take encouragement for where I was in 2009 and where I ended up today.

And so again, I dedicate this book (and all my previous books) to those who follow in my path building and leading project communities all over the world, sharing knowledge and passion and inspiration to the project world, and just loving our project profession – the world needs us, well it certainly needs you.

The future is yours, not mine. I just helped out a bit in the last few years!
Peter Taylor (aka The Lazy Project Manager)

Note

1 My official 500th presentation took place in Budapest, Hungary, at the excellent PMI event in November 2023.

Index

Printed in the United States
by Baker & Taylor Publisher Services